THE MANAGERIAL IMPLICATIONS OF MICROELECTRONICS

Microelectronics and microprocessors have only recently become familiar words, frequently associated with depressing implications. The British Government has taken the unprecedented step of investing substantial sums to make industry 'aware' of this new technology.

Nevertheless most managers remain baffled. Whilst aware of the significance of this new technology they find it difficult to assess how far-reaching its effects will be. In particular, managers are concerned to know what they should do about it individually and when.

This book is written to help the manager clarify some of the questions that have arisen through the development of this new technology. Bringing together a distinguished group of authors who have been studying these problems from a variety of viewpoints – from within industry, trade unions, universities and consultancy – it examines the strategic implications, the effect on employment and how the unions might respond, the impact in production processes and systems, office procedures and employment, etc. More importantly, the wider managerial consequences are stressed throughout the book. The immediate problems may be technical but the future will be more concerned with the social, organisational and broader managerial aspects.

It is too soon to see clearly all the managerial implications of microelectronics, but this book provides managers with a valuable focus for their thoughts.

Also by Brian C. Twiss

MANAGING TECHNOLOGICAL INNOVATION

ORGANISATIONAL PROBLEMS IN EUROPEAN MANUFACTURE (vols 1 and 2) (*with T. D. Weinshall*)

FORECASTING TECHNOLOGY FOR PLANNING DECISIONS (*with H. Jones*)

MANAGING INDUSTRIAL ORGANIZATIONS (*with T. D. Weinshall*)

SOCIAL FORECASTING FOR COMPANY PLANNING (*editor*)

THE MANAGERIAL IMPLICATIONS OF MICROELECTRONICS

Edited by
Brian C. Twiss

First published 1981 by
THE MACMILLAN PRESS LTD
London and Basingstoke
Companies and representatives throughout the world

ISBN 0 333 28090 3

Printed in Hong Kong

For
Pam, Martin and Catherine

Contents

List of Figures

List of Tables

Notes on the Contributors

Brian C. Twiss

Assistant Director of Management Development Programmes and Head of the Bradford Group for Forecasting and Policy Alternatives. Before joining Bradford he was Director of the Cranfield Management Development Programme and Head of the Case Research Project. Previously he worked for a number of years in project management in advanced aerospace research and development in the UK and Australia. He has consulted and taught on senior management programmes for many large technological companies, business schools, and government research institutes in Britain, the Continent of Europe, the USA, Israel, and South Africa and presented papers on aspects of R&D management at conferences in the UK, Paris, and Washington. He has published widely on the subject of technology management and forecasting for technology and his writings have appeared in Italian, Romanian, and Spanish translations. He is author of *Managing Technological Innovation* (Longmans, 1974), co-author of *Organisational Problems in European Manufacture* (Longmans, 1973), *Forecasting Technology for Planning Decisions* (Macmillan, 1978) and *Managing Industrial Organizations* (Pitman, 1980) and editor of *Social Forecasting for Company Planning* (Macmillan, 1981).

Paul F. Bestwick

Started as a craft apprentice with Crofts (Engineers) Ltd, where he gained a wide range of experience during his 11 years with this organisation. In September 1969 he became a lecturer in Operational Research and Statistics at Leeds Polytechnic Department of Management and Business Studies. He joined the University of Bradford Management Centre in January 1973 where he is Senior Lecturer in Production Management. He has contributed to leading management journals and is co-author of two books on Production Management published by Pitman early in 1981. He has acted as a consultant to local and multinational industry.

Ray Curnow

Ray Curnow is Director of Probit Consultancies Ltd, a consultancy concerned with the economic and social impact of technological change. From 1966 to 1979 he was a Senior Fellow of the Science Policy Research Unit. His recent work includes studies of the social, economic, and industrial impact of information technology underpinned by microelectronics, optical fibres, and satellites. His publications include *The Future of Micro-electronics: Forecasting the Effects of Information Technology* (Francis Pinter, 1979), written with James Barron of INMOS. He is an Associate Fellow of the Institute of Education, University of London and a Visiting Honorary Fellow to the Bradford Group of Forecasting and Policy Alternatives at the University of Bradford.

Denis C. Dalton

Until recently Director of Technical Services of The Plessey Company Ltd, he joined Plessey in 1952 and until 1970 held a number of appointments in R&D concerned with design and production engineering of communication navigation and instrumentation products. For 8 years he carried sole responsibility for all the activities of 1600 technical staff during the period when miniaturisation and subsequently microelectronics was developing. From 1970 to 1974, Mr Dalton saw widening experience in general management including significant spells in personnel management, works management, transport and similar fields. From 1974 as Corporate Staff member for technology, he advised the most senior company management on product policy, technological forecasting, project and technical management. During this time he paid particular attention to the accelerating impact of microelectronics on products, people, and processes.

Peter G. Hardie-Bick

After training and serving in the aircraft industry mainly in production management Peter Hardie-Bick joined the PE consulting group where his work was mainly in the planning and control of engineering production. He joined the central management advisory unit of Unilever in 1963 where the main part of his work was in connection with the development of coordinated systems of planning and controlling manufacturing sales and distribution of consumer products. He was seconded to the Malawi Government through the ODM as General Manager of the Malawi Development Corporation for several years where his main interest was in finding small-scale solutions to many familiar manufacturing problems. Latterly he has been involved in the practical application of microprocessors to management problems within Unilever and elsewhere, specialising in the use of programmable calculators as an alternative to more sophisticated solutions.

Joseph J. Hunter

Dr Hunter is a Principal Scientific Officer in the Automation Division of the National Engineering Laboratory. He has extensive experience in the application of electronics to engineering research. In recent years his research has been concerned with the automation of arc welding and the development and application of industrial robots and microcomputers in engineering. Author of many papers on the subject of electronics, instrumentation and automation he also holds six patents.

Frank Jones

Frank Jones, FMS, MBIM, MInstAM, AIDPM, MIRT, is the Managing Director of Business Information Techniques, a company providing consultancy on a wide range of office automation topics. He pioneered the use of word processing in local government, installing stand-alone systems at Cheshire County Council in the early 1970s and screen-based shared-logic systems at the City of Bradford Metropolitan Council in the mid-1970s. He is a former contributor to the Prime Minister's Think Tank on Microprocessors, co-author of the LAMSAC report on Word Processing, and a regular contributor to the technical press. He has been involved most recently in developing and implementing strategic office automation policies for several multinational companies.

L. March

Leonard March, BSc, C Eng, MIERE, has spent 6 years on research and development of instrumentation and telemetering equipment. He subsequently joined IBM laboratories and worked on processor and memory device development. Since 1977 he has been with McKinsey & Co. Inc., working with clients throughout Europe on the problems of developing and implementing technological strategies and of managing industrial research and development.

A. E. Owen

Tony Owen has to date 15 years' industrial experience in the special-purpose machine tools and high-technology instrument industries, both in the UK and abroad. His last post abroad was as General Manager of a high-technology company. He successfully completed his MBA at the

University of Bradford Management Centre in 1979 where he specialised in Production Management. Much of his chapter is based on his MBA dissertation. Currently he is Project Manager at Remek Automation who manufacture special-purpose machine tools and are developing robotics.

Barrie Sherman

Director of Research, Association of Scientific, Technical and Managerial Staff (ASTMS). A qualified dental surgeon, he read economics at Queen Mary College (University of London) and at the University of California. Member of the Wilson Committee on Financial Institutions and three subcommittees of the Labour Party National Executive. Author of many articles and co-author with Clive Jenkins of four books, including *The Collapse of Work* (Methuen, 1979).

Peter Trier

Dr Trier graduated as a Mathematical Wrangler at Cambridge, and went into the Royal Naval Scientific Service. In 1950 he joined the Mullard Research Laboratories where he was Director from 1953 to 1969. He then became Technical Director of Philips Industries, London. In addition, Dr Trier is currently Chairman of Council of Brunel University, and Chairman of the Electronics Research Council of the Ministry of Defence. He has contributed to many technical books and journals, and has lectured widely on technical and industrial topics. Recent lectures include the British Association, both in 1976 and 1977. He was awarded the CBE in the January 1980 Honours List.

M. F. Woods

Dr Woods graduated from London University with a PhD in Pure Chemistry in 1957 and moved from bench work to Process Design and Plant Commissioning, first with the National Coal Board in the UK, and then with the Unilever Food Group. His interest in Management and OR systems took him away from the functional side of the business and progressively into internal consultancy for Unilever (UK) Holdings Ltd and then independently. He is currently Assistant Director of the Management Development Programmes at Bradford Business School, where he is responsible for the Production Management Programme. His special interests are the organisational changes in business, and in particular the changes being brought about by the microcomputer. He has several patents and articles on a variety of subjects to his name.

Introduction

The history of industrial society has been transformed by a few major technological events. The harnessing of steam power provided the stimulus for the first industrial revolution. Our own society has been largely shaped by the application of electricity and the internal combustion engine. Now we are witnessing the beginnings of another era based upon microelectronics. Further ahead we see the prospects of biotechnology.

The major impact on industries, managements and society takes some time to manifest itself. For what may sometimes be a long period the new technology is costly and unreliable. Thus it can be dismissed as a useful contribution to economic wealth but by no means a revolutionary force which will have far-ranging effects. Microelectronics is no exception. It is now 30 years since the first elementary computers were designed. The increasing applications of automation have a long history; but as will be seen later this is a characteristic of all technological developments. After a period of slow growth often extending over several decades there comes a time of rapid exponential growth heralding widespread impacts. This is when many existing businesses have to undergo profound changes if they are to survive or meet the competition of new businesses which have grown up with the new technology.

It would be wrong to confine one's attention to the technological aspects. Admittedly there are major technical problems in applying microelectronics, particularly in the short term. The main long-term effects are, however, more far-reaching. In some cases the product base of the company will be threatened. In almost every industry we shall see profound changes in marketing, production, the managerial information systems, the role of the individual manager, the quality and quantity of all employees including managers and the organisational structure of the firm. Nor can any consideration of microelectronics ignore the impact upon the society within which the company operates, for this in turn is an inescapable element of the organisation's strategic context.

Many managers may wish to defer consideration of these matters, coming as they do when they are beset with short-term problems stemming from the economic difficulties currently facing them. This new technology, they may think, could not have arrived at a worse time in industrial history.

A major deterrent is uncertainty and fear of the unknown. From the writer's own discussions with industrial managers it is clear that they feel threatened by a technology which is beyond their experience although becoming familiar to today's generation of schoolchildren. Some schools already possess more minicomputers than some quite sizeable companies. This fear is largely unwarranted for the manager does not need a detailed knowledge of the technology in order to manage it. His concern must be with the managerial implications.

This book is intended for the general manager. It explores these wider implications and raises some of the issues to which he should turn his attention. It cannot, of course, answer his specific queries for these will vary between industries and between companies. It should, however, become clear to the reader that this is a technological development from which no manager can escape.

Chapter 1 gives a brief survey of the total picture and introduces topics which will be explored in greater depth in subsequent chapters.

Chapter 2 is a simple explanation of the main features of microelectronics, written in terms the layman can understand. It focuses on what they can and cannot do. This chapter can be ignored by those readers who already possess a knowledge of electronics or computers.

Chapter 3 examines some of the strategic implications. It emphasises the rapidity of the pace of progress in recent years and indicates some of the business areas where the effect is likely to be most marked.

Chapters 4 and 5 deal with the important topic of employment. Much public debate has been focused on the prospects of long-term unemployment resulting from the widespread adoption of microelectronics. Although opinions are divided it appears to most contributors to this book that there will be a significant reduction in the number of employees both in the office and in manufacturing. This will have significant social and political consequences. How, for example, will the trade unions respond? It is difficult to answer this question since union views are divided. A valuable insight into the considerations which will influence their response is given by a leading trade union thinker in Chapter 5.

The next two Chapters, 6 and 7, deal with production, the first examining the development of industrial robots–the use of which has recently been expanding rapidly in Japan and the USA but at a much slower rate in the UK. Equally important is the application of microelectronics in the management of production. This is discussed in Chapter 7 which also describes a number of recent applications and the lessons to be learnt from them.

Chapter 8 looks at management systems and some of the ways in which the role of the manager in relation to the total system will be modified.

Chapter 9 explores some of the effects upon the office, and studies a detailed case of the successful introduction of word processors in a local authority.

Chapter 10 is concerned with service industry. In particular the reader will see how many of the managerial techniques which have not been widely used in the past can now be used by managers with an ability to use simple programmes and a microcomputer.

Finally in Chapter 11 the experiences of a major international consultancy in the implementation of micro-electronic change programmes will be described.

This book makes no claims to be a comprehensive coverage of all the managerial impacts of a technology so all-pervasive as microelectronics. To do so it would become either excessively lengthy or unduly superficial. It does, however, raise many of the major managerial issues, and it discusses the more important of them in some depth. It will have achieved its aim if it encourages the reader to accept the inevitability of the changes which will be brought about by the microelectronic revolution and to start planning for their introduction in his own organisation.

1 Microelectronics – The Managerial Dilemma

BRIAN C. TWISS

Management's prime role is to ensure the long-term survival and profitability of their businesses. Today's managers' attitudes have been formed by their experience of post-war economic growth and the steady development of well-established technologies. In the 1980s, however, the pace of economic growth already slowing in the late 1970s will continue to slacken; some forecasters predict a decline with unemployment rising even without the impact of new technologies. On top of these economic uncertainties managers now have to face up to the dramatic effects of the microelectronic revolution, heralding what has variously been called the Third Industrial Revolution, the Information Society, or the Post-Industrial Society. This is a new world for which the manager has little training or psychological preparation.

Bombarded by reports, conferences and the media, few managers can be unaware that something momentous is afoot. To make matters more confusing there are widely differing forecasts emanating from the many institutions, both public and private, claiming to speak authoritatively on the subject. One example is provided by the prospects for employment. Whereas some commentators foresee manufacturing employment falling rapidly to a level comparable with agriculture, others suggest that new technology-based, innovative products and services will provide unforeseen job opportunities.

Every manager has to take a view of the future. However difficult this may be he has a responsibility to study carefully the trends that are evolving in the world so that his decisions can reflect his judgement of how they will affect the future of his business. In times of great uncertainty it is asking too much to expect accurate forecasts and optimum decisions. Nevertheless he should reduce the risks to which he is exposing his organisation and increase the opportunities he identifies and exploits. This is the approach that should be adopted towards the application of microelectronic technology. For at this relatively early stage of the diffusion of this technology into business, managers cannot expect to obtain clear and precise answers to the questions they should be asking; namely:

(a) *What effects* will microelectronics have on my business?
(b) *When* will these effects begin to become important?
(c) At what *rate* will their impact grow?
(d) What *action* should I take now?

In spite of the difficulties inherent in answering these questions it should be possible to identify those areas where major threats and opportunities may be imminent; where they may emerge in the medium term, say 5–10 years; and equally important, where they are likely to have little impact.

THE DIFFUSION OF TECHNOLOGY

There is no reason to expect microelectronics to exhibit a different growth pattern from technologies of the past. This is characterised by an S-shaped growth curve (Figure 1.1). From this it can be seen that the initial period of growth is slow; much slower than the technological optimists might suggest. This is followed by a rapid rise which slows again as saturation is approached. Each application of microelectronics can be expected to have its own S-curve. The electronic watch, for example, is already well advanced up the curve; but for most applications the point of take-off has not yet been reached. The problem facing the businessman is well illustrated by the example of the fully automated factory. Articles appeared in management journals as long ago as the late 1950s predicting the imminence of the fully automated factory, but no dramatic changes occurred in the following quarter century in spite of the increasing use of transfer lines and numerically controlled machine tools. The ultimate 'end-state' of the fully automated factory still seems a reasonable assumption. But when will it occur? Unfortunately much of what has been written regarding microelectronics is of this nature, describing scenarios of the ultimate situation without giving guidance about the path by which it will be achieved. This is not of much value to the decision-maker in industry who is concerned with the actions he needs to take now or in the near future. He may be interested in scenarios for his industry 50 years hence but they are of little value if they are unlikely to influence his current

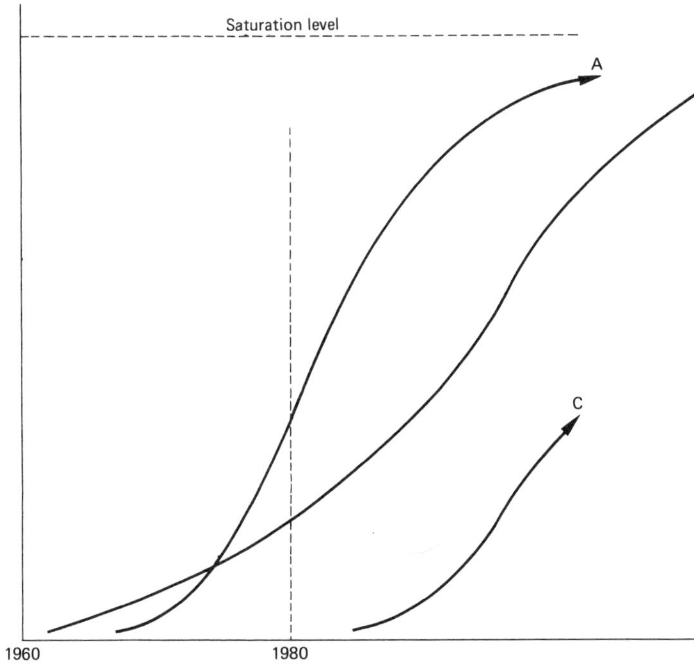

FIGURE 1.1 Microelectronics and the Growth curve

NOTE: Growth varies in timing and rate. Curves A, B, and C show the different curves for:

A electronic watches;
B industrial robots;
C fully automated mining machinery.

These curves are illustrative and do not purport to be accurate representations of the growth of these three applications of microelectronics

decisions. Returning to the case of the automated factory we may conclude that the recent growth of industrial robots does mark the turning point for many manufacturing industries and deserves active consideration now, some 25 years after the initial optimism.

It is useful to consider what determines the shape of the S-curve. Usually a new technology is expensive, because it has not yet benefited from learning curve effects, and is unreliable. (Research into many industries shows that costs fall exponentially with accumulative production volume, frequently with a slope of about 70 per cent; i.e. the second unit costs 70 per cent of the first, the fourth 70 per cent of the second, the eighth 70 per cent of the fourth, and so on (Hedley, 1976). Thus its initial impact is likely to be where its technical features are suited to a specialist application where high cost and relatively low reliability are acceptable. As experience accumulates cost falls and reliability improves to a point where it becomes competitive in high-volume applications. Microelectronics is remarkable in its extremely rapid fall in price; thus it has been adopted widely where the facility represented by the electronic components represents a substantial proportion of the total cost of the item, hence the rapid growth in watches and measuring devices. Where the micro-processor is a smaller proportion of the total cost of a system involving sensors and control mechanisms such as industrial robots, the growth rate has been slower.

These considerations suggest that the manager might pose the following questions:

(1) Where might microelectronics enable me to perform desirable functions which have not been feasible in the past:
   (a) where the microelectronics are the major component?
   (b) where the microelectronics are part of a larger system?
(2) When will microelectronic systems fall to a price sufficiently low to be competitive with current systems employing other technologies? This gives an indication of the time when the S-curve will begin to rise sharply.
(3) How fast will the price of microelectronic systems continue to fall? This will give an indication of the slope of the rapid growth phase of the S-curve.
(4) What performance levels are required from other parts of a system (e.g. sensors) to make a microelectronic system:
   (a) feasible?
   (b) competitive?
(5) What is the main advantage to be gained from the first application?
   lower product cost;
   improved product performance;
   improved product quality/reliability;
   elimination of dirty, unhealthy, hazardous jobs;
   overcoming labour shortages – skills or availability;
   provision of better managerial information and control.

These questions are based upon techno-economic considerations. Increasingly, however, social and political factors are becoming an important element in managerial decision-making and need to be considered carefully when assessing the merits of microelectronic applications. In some instances they will provide powerful support for the early adoption of the new technology. For example it has been the cost, quality and availability of clerical staff in city centres which have provided the main motive for the use of word processors by a number of financial institutions. A major incentive for the adoption of computer-aided design has been the shortage of skilled drawing office staff (ACARD, 1980). Similarly robot applications have been concentrated on dirty or hazardous tasks such as paint-spraying. In other cases, however, the installation of microelectronic systems is likely to be delayed by employee or trade union resistance where large numbers of jobs appear threatened. Any analysis of the merits of microelectronics which does not consider these social effects explicitly is likely to lead to poor managerial decisions.

TIMING

Even where future applications may appear clear the manager must still decide when and on what scale he should take action. Should he allow competitors to pioneer the system and learn from their experience?, or should he attempt to lead?

One feature of new technology, all technology, is that it usually means trouble (Twiss, 1980). Many problems will occur which could not have been foreseen however careful the planning. Furthermore, the initial benefits are likely to be substantially offset by additional costs not taken into account in the original computations of benefit : cost. This should not be regarded as a good reason for delay; rather the reverse. One cannot easily benefit from climbing on the back of someone else's learning curve. However well-developed the system might be there is no substitute for learning from one's own mistakes and experience. Microelectronics, as will be seen in later chapters, is likely to have wide-ranging effects throughout the business, technically, organisationally, in managerial systems and work practices. The changes involved take time. They should not be introduced where their initial success is critical and where time scales are tight.

Thus it would appear that the answer to the question 'when?', should be 'now'. For a technology which will have such a wide-ranging influence throughout all types of business needs to be assessed in the light of one's own experience. This should be regarded as an investment for the future. The less one knows about the technology the more vital it is to gain some limited operational experience. This enables the problems to be tackled while they are still minor and whilst there is ample time to experiment and find adequate solutions.

WHY MICROELECTRONICS ARE IMPORTANT

The characteristics of microelectronic technology will be explored at greater length in subsequent chapters. It is useful, however, to summarize the main features which underline their immense potential:

(a) *Cheapness.* The dramatic fall in price per bit from $2 in 1970 to $0.02 today is illustrated in Table 3.4. Although this fall will continue the low price of the basic chip is such that it can be virtually neglected in most applications. The cost of a system is now essentially that of the other components (e.g. sensors) and the software.

(b) *Reliability.* The microelectronic circuits are inherently reliable when compared with the systems they replace. They do not, for example, employ moving parts as in electromechanical systems which are prone to failure and require routine maintenance in addition to fault rectification. This reliability is, of course, a mixed blessing. For the user it brings substantial savings. The producer, however, has to weigh up a variety of factors:
   (i) reliability brings extended product lives, thereby reducing the size of the replacement market and the demand for spares often more profitable than the initial sales;
   (ii) cheapness mean a lower financial turnover unless the price-elasticity of the product is such that it can be compensated by an increased sales volume.

(c) *Small size.* The physical size of a microprocessor is very small when compared with the components it replaces. In some applications (e.g. the electronic watch) this will have little effect upon the size of the final product. But in other uses the impact can be substantial (e.g. electronic telephone exchanges). In these latter cases there can be substantial savings in the size of buildings, space-heating systems, etc. This in turn will indirectly affect the building industry.

(d) *Speed of operation.* A microelectronic system operates almost instantaneously. Mechanical and electromechanical systems are largely limited by inertial effects. The elimination of these inertial limitations enables electronically controlled equipment to operate at higher speeds; thus, in the absence of other constraints the same output can be achieved with fewer machines. As seen already this can result in a reduction in physical facilities.

(e) *Accuracy.* A digital system does not deteriorate in performance unlike an analogue system which usually suffers from wear. Thus a microelectronic system can be set up accurately and will continue to operate as designed until it ultimately fails. In contrast the performance of a mechanical control system (e.g. a cam) will deteriorate with time. It will also be noted that a cam requires lubrication. Thus the electronic replacement may eliminate the necessity for a lubrication system with all its attendant problems.

(f) *Low energy consumption.* The energy consumed by a microprocessor is negligible. Remembering that the technologies it replaces are also information processing systems the scope for energy saving is strictly limited. The main advantage lies with its ability to operate for long periods with a small power source. Thus the low consumption enables an electronic watch to operate for three or more years from a small battery, whereas earlier electromechanical watches were bulky and required frequent battery replacement.

Although there will be little energy saving from the information system itself, the microprocessor can contribute to energy conservation within the power system in which it operates through more efficient control. Thus in the motor car economy will be achieved through improved fuel and ignition control, and in the home through more efficient management of central heating and domestic appliances.

(g) *Low material usage.* The chip itself uses only small quantities of a cheap material silicon. It replaces thousands of other components connected by wires or printed boards. This does save some material but not a significant amount. The main saving is indirect – small equipment racks, higher productivity from machinery hence fewer machines, smaller buildings, etc.

(h) *Safety and environmental acceptability.* Most technologies have undesirable side-effects: atmospheric pollution, nuclear hazards, the unknowns surrounding biotechnology. Microelectronics is almost unique in being entirely free from safety or environmental objections. Indeed concern for the environment favours certain applications, for example, where improved combustion and emission control can be obtained using microelectronics, or where it enables the human operator to be removed from a hostile working environment.

The characteristics we have examined, whilst not exhaustive, give some idea of the scope of the potential benefits. On economic and technological criteria there are no disadvantages to offset the gains. Additionally we have noted a few of the areas where there are secondary advantages. Economists refer to the 'multiplier effect' whereby demand for a product such as the motor car creates demands in the energy industry (for fuel), metal industry (for the materials), the extractive industries (for the raw materials), the building industry (for garages) and so on. By contrast microelectronics has the opposite effect by lowering demand in many associated activities. This is one of the reasons why its impact on employment is likely to be significant.

It must be stressed, however, that we have been discussing the effects of the basic unit – the chip. Much of this potential has yet to be released, and to do so one has to consider the total system into which it is embodied including the software, and, of course, any social consequences of a particular application.

THE IMPACT ON BUSINESS

All businesses will be affected by microelectronics; but the significance for any individual firm will depend upon the character of the industry in which it is engaged. We shall now briefly examine some of these influences under the following headings:

(1) Indirect effects through society.
(2) Strategic.
(3) Degree of industrial integration.
(4) New products.
(5) Production.
(6) Marketing.
(7) Research and development.
(8) Personnel and management.

It must be recognised that it is too early to see clearly what all the effects may be. For this reason readers may not agree with all the conclusions drawn. Nevertheless the purpose of this introductory chapter will have been served if it directs attention to some of the issues of importance.

(1) INDIRECT EFFECTS THROUGH SOCIETY

No business can divorce itself from the society within which it is operating. It must respond to the needs of that society and the pressures emanating from it.

A great deal has been written about the relationship between microelectronics and employment, some of the arguments will be repeated in later chapters. The factors to be weighed are:

(a) labour replacement in current manufacturing operations *v.* job creation for the manufacture of new products;
(b) labour replacement in offices and service activities *v.* the creation of new jobs in offices and services;
(c) international competitiveness which will determine the balance between manufacture within developed nations and between the developed and developing nations.

The writer's view is that the balance of these factors will lead to a significant displacement of employment. If this view is wrong it will only be invalidated by a substantial change in working conditions, the type of employment and employee attitudes. In either case it will produce powerful forces in the societal environment within which every business operates.

A number of scenarios have been published depicting possible future societies. It is beyond our scope to discuss these in this book but this does not mean that senior management should not be studying them seriously. As an illustration we might examine a society where there is a high degree of work-sharing (i.e. shorter working week and longer holidays) and an increase in unrest amongst the unemployed youth (e.g. vandalism and violence). This would affect all businesses whether or not they had a direct application for microelectronics. A labour-intensive operation would still have to match the working conditions of other companies in the area; it could affect decisions on the siting of new factories nationally or internationally; and could necessitate a re-examination of security arrangements.

2. STRATEGIC IMPACTS

The greatest danger to which a manufacturing company can be exposed is the threat to its product base. Industrial history is full of examples where new technologies have destroyed complete industries. Microelectronics has already added to this catalogue of disaster. Slide-rules have disappeared, electronic calculators have displaced the electromechanical machine and some of the manufacturers have gone out of business, and the Swiss watch industry has suffered a dramatic setback. No manufacturer of electromechanical devices is immune. So much is already clear, but can management make the necessary adjustments? How strong is the 'It couldn't happen to me' syndrome? Even when the need is recognised can the expertise in an entirely new technology be acquired?

Frequently the new technology is disregarded for what appear at the time to be sound reasons. It is seen to be unreliable, a characteristic of all new technologies, but the rapid improvements to be expected are not sufficiently anticipated. The economic benefits may also appear dubious, although these will increase rapidly due to the 'learning curve' effect discussed earlier. Thus there is a natural inclination to under-estimate the threat. The further the new technology is removed from current practice the greater this risk is likely to be. These dangers are likely to be more acute in a technology which is advancing at the speed of microelectronics. This does not, of course, mean that management should rush into the new technology regardless. But it does imply that it should be examined realistically and sympathetically where there is a possibility of a major strategic threat.

Which are the industries under possible threat from microelectronics? No easy answers can be given since a detailed assessment and forecast incorporating all the technological, economic, and social factors involved would be required for each industry. As a generalisation one can start from the fact that microelectronics is essentially an information processing and communications medium.

Many of the potential applications relate to the transmission of data and other information. Thus it would appear that facsimile transmission could replace much of the material carried by post and the printed newspaper. Why use paper when electronic storage and transmission could be utilised? The major determinant here could be the rate at which human attitudes can change rather than the economic and technical feasibility. For example, it might be argued that paper provides a more permanent record for important documents such as the deeds for property. But is this really so? It is amusing to speculate on the objections that might have been raised to the introduction of papyrus to replace the indestructible record provided by stone!

The difficulties facing the forecaster in this area can be illustrated by the example of the video-telephone which it was claimed would enable the commuter to work from home, rather than to move physically to his place of work. These forecasts made 15 years ago have yet to come about. And yet today it is confidently predicted that much trans-Atlantic business travel will be replaced by satellite transmitted tele-conferencing facilities by 1990.

These examples highlight the difficulty in assessing some of the strategic threats. One can state confidently that a number of industries existing today will be partially or wholly replaced by microelectronic applications. It is less easy to forecast which, and the timing. Yet the possible dramatic impacts are such that senior managements cannot afford to ignore them.

3. DEGREE OF INDUSTRIAL INTEGRATION

Most manufacturers are both makers of their own parts and assemblers of parts bought from other manufacturers. Apart from a few specialist electronic manufacturers the heart of any microelectronic system, the chip itself, will be bought from one of the few companies mass-producing them. This will substitute for a variety of electrical and mechanical components formerly manufactured by the maker of the end product. A similar trend might be expected to emerge for sensors and other key elements of the final system.

It is likely, therefore, that many companies will become more reliant upon their assembly operations. This is the opposite of the trend towards vertical integration which has been evident in many industries over the past 20 years. It will consequently reduce their value added in production. Many makers of electronic equipment (e.g. control systems) are already essentially assemblers of components bought from a number of international sources. The characteristics of microelectronics are such that they are likely to lead to a spread of this trend into other industries in the years ahead, and provide many opportunities for the founding of new small businesses.

On the other hand the increasing capital intensity of manufacturing will favour the large company with the benefit of economies of scale leading to a concentration into a few large companies (e.g. motor manufacture). Thus the effects of the microelectronic revolution must be considered in relation to the characteristics of the industry; in some cases it will favour the proliferation of small businesses, in others it will lead to amalgamations and a concentration in a few very large organisations.

4. NEW PRODUCTS

It is beyond our scope here to catalogue all the feasible new products based upon microelectronics. Indeed one of the major concerns of manufacturers is to decide which of the many potential applications should be supported. Some applications in specific areas will be discussed in later chapters.

It is worth, however, examining briefly the nature of new products. The most important fall into three categories:

(a) those substituting for products currently manufactured;
(b) those substituting for products of other manufacturers, frequently using other technologies;
(c) entirely new products.

The majority of new products fall into the first category. They will succeed in the market place either because they are cheaper or because they offer a higher performance; in many cases there will be a combination of these two factors. Thus in TV we have seen the substitution of microelectronics in monochrome sets leading to a lowering of their real cost. At the same time there has been a substitution of colour TV for monochrome as the main viewing set. In the next decade we shall witness the substitution of the flat screen for the conventional tube.

In most cases the long-term trend in price (in real terms) is downwards. Thus if the new products do not lead to an increased level of ownership the impact on the industry can be a substantial fall in turnover. This is likely to be the effect on many products incorporating microelectronics. In assessing the future for new products it is therefore necessary to evaluate the technological potential for a stream of improved products to maintain market demand against the degree of penetration for the product type. For example, an improved razor blade is unlikely to increase the number of people who shave or the frequency with which they shave.

We have seen when discussing strategic effects that a major result of microelectronic-based products will be to substitute for other products incorporating other technologies. Thus the electronic calculator has replaced the slide-rule, log-tables and the electromechanical desk calculator. Although this has resulted in a vast growth in the ownership of such devices it is claimed that the sales value of pocket calculators and the employment in their manufacture is little different from that of the old desk top calculator. This 'innovation by invasion' from the substituting electronically based equipment provides many profitable opportunities for the electronic industry but raises major threats to the products of non-electronic industries. Moreover, although the level of ownership may be greatly increased, there is unlikely to be a significant growth in turnover of industry as a whole. Indeed the reverse will often be the case. It is also difficult to avoid the conclusion that in almost every application the level of employment will fall.

The future growth of the electronic industry will be based upon entirely new products and electronic systems – TV games, video-recording and Prestel are but three current examples. The potential number of new products is immense. When, however, one considers the new products, it will be found that in almost all cases they indirectly substitute for an existing product or human task. For example, let us take automated check-out equipment in supermarkets. This may lead directly to the need for fewer check-out operators; but it also leads to more efficient stock control thereby reducing the need for warehousing and waste through time-expired goods. Similarly the TV game has probably affected the sale of board games.

It is difficult to generalise about the future effects of the wide diversity of new products which will be based upon microelectronics. It is even more difficult to be specific at this early stage of their development. Nevertheless, it can be seen that they are likely to have widespread impacts on almost every other industry and area of business activity. Apart from the electronic industry itself, it will lead to a fall in the demand for materials, non-electronic products and labour.

5. PRODUCTION

Microelectronic applications in production have received a great deal of publicity particularly in view of their impact in the socially sensitive area of employment. Industrial robots have taken the limelight but other applications in the control of production operations and in their management are of equal if not greater importance. Both these aspects will be discussed in detail in later chapters, so we shall confine ourselves in this introduction to a brief summary.

Robots can be regarded as the ultimate stage in the process of automation that has been advancing steadily for over 40 years. They have now reached a stage where their adoption is likely to grow rapidly and their uses to expand from their current applications in paint spraying, welding and simple assembly operations. Outside manufacturing industry it is possible to foresee widespread use in mining and deep-sea applications. At present many possible uses are limited by insufficient development of the sensor system, a problem which will be gradually overcome with development; the unreliability which is to be expected in the early stages of a complex technology; and opposition aroused due to fears of unemployment. For the next few years the technical problems are likely to prove the main limiting factor, but this is likely to shift to social concerns as they become more widespread.

There are, however, few areas where microelectronics cannot be applied to the control of production processes. It is in the simpler applications that they have often made the greatest impact so far. For example, the author recently saw an application where a simple microelectronic control circuit was used to replace the operator of a plastic blow-moulding machine; the payback period for this application was 2 weeks. The main advantages of these systems are:

(a) higher production rates, largely through the removal of human constraints (no tea-breaks, overtime rates, strikes, etc.);
(b) accurate control of the operation;
(c) accurate reproduction thus enabling the elimination of separate batch identities where technical limitations now make this necessary – this is an advantage to the consumer as well as the producer in reducing inventories and the need to identify batches;
(d) the ability to change products on the production line rapidly – this applies also to continuous operations such as carpet manufacture (where short lengths of a pattern can now be produced without a significant cost penalty) and the weaving of clothing name tapes.

Whilst the technical problems of introducing the new systems may be the most immediate, it is the longer-term effects to which managers should be turning their attention. In most cases the introduction of the new systems will be gradual. It is thus easy to overlook the fundamental changes which are taking place in their organisation. A few examples will be given to illustrate the trends which are likely to occur.

As the number of employees falls so will the company become more capital-intensive. Thus many manufacturing companies will grow increasingly like process industries. They will become less flexible in their response to changing economic conditions. The emphasis will shift from labour productivity to capital equipment productivity and the need for continuous operation throughout the 24 hours of the day and throughout the year. If the current practice in process industries is taken as a guideline they will find it necessary to engage in more formal planning and long-term forecasting.

The nature of jobs will also change. There is likely to be a polarisation between the highly skilled and the unskilled. The highly educated machine-watchers will be idle for long periods interrupted by periods when they will be using all their practical and intellectual skills to cope with a malfunction. At the same time it will not be economical to eliminate many of the most menial tasks. Thus there is likely to be a polarisation between the highly skilled and the unskilled with very few semi-skilled. How will this affect the social system within the company? What effect will this have on the style of management and the type of manager?

Much labour reduction is currently achieved by natural wastage. As men retire they are not replaced by young recruits. Thus the average age of the work-force increases. Apart from the social consequences of youth unemployment such policies make the transition to the new technology more difficult, for it is the youth who will have the skills and education appropriate to the new systems. Alternatively it may be argued that it will be necessary to retrain many of the existing work-force. Can it be done? Will they be willing to undergo retraining?

The requirements for skills in servicing and maintaining much production equipment may also change. In the past electrical, hydraulic, and mechanical units were sufficiently discrete to be related directly with workers possessing only one of the skills. Increasingly these equipment demarcations are disappearing, indicating a need for multi-disciplinary individuals rather than multi-disciplinary teams. How should these people be trained? What effect will it have on the existing social and pay differentials? Will it be acceptable to the trade unions? These and many other considerations need to be examined carefully by management if they are to manage the transition smoothly.

6. MARKETING

Traditionally there have been many causes of conflict between marketing and manufacturing due to the inability of the latter to respond rapidly and economically to the varied requirements of customers. The application of microelectronic technology to both design and production enables many of these difficulties to be reduced.

Computer-aided design now enables products to be modified easily and quickly. Thus the penalties for design changes are considerably reduced thereby permitting:

(a) the early modification of products to incorporate the latest technology;
(b) the rapid design of new products to meet changing market requirements;
(c) the economic design of products tailored to the demands of individual market segments or customers.

Although constraints on the extent to which these advantages can be exploited in practice will not be completely removed, they are nevertheless likely to lead to increased product variety and shorter product lives.

These benefits could not, of course, be realised unless production could respond to the demand for frequent changes both in relation to design modifications and the product mix. But this will become increasingly the case due to the application of microelectronic systems in both production equipment and production control methods.

Thus the combination of computer-aided design and the increasing flexibility of production equipment will enable the manufacturers to satisfy the needs of the customer more closely. There will, of course, be an associated managerial penalty from this added complexity. Yet, here again, the improved information systems will alleviate this problem.

We have seen already, however, that the individual product is likely to last longer due to the improved reliability associated with microelectronic systems. Thus we have two conflicting forces at work. It will become easier to make a product obsolescent but at the same time the item once purchased will last longer. The balance of advantage to the manufacturer will depend to a great extent upon the technological potential for development of the product. Where this is substantial the customer will replace it more frequently, thereby increasing the demand; where it is slight the customer will be less inclined to purchase a new item due to the enhanced longevity of the one he already possesses, thereby reducing overall demand.

Improved information systems will enable the provision of a better service to the customer in a variety of ways. Information on demand fluctuations will be available immediately and production will be able to respond more fluidly. The results of special promotions will be evaluated more quickly. Information on the quantity and location of items will be available on demand. Much of this is already with us, but usually it is confined to large suppliers with expensive computer installations. In the future this will be the accepted norm throughout industry.

The benefits will not only be derived through speed of response. A major cost in all organisations, though difficult to quantify, stems from wastage of effort due to poor information, dealing with customer enquiries or complaints, tracking down spare parts, progress-chasing, etc. The improved information systems will enable the elimination of much of this wasted effort.

7. RESEARCH AND DEVELOPMENT

The development of our industrial society can be thought of as consisting of three periods. Before the industrial revolution the cost of a product consisted of labour and material. The industrial revolution, although based upon a few key inventions, was mainly characterised by the addition of capital. In the past 50 years the contribution of organised intellectual effort through research and development has become of increasing importance. Thus we now have four key resources: labour, materials, capital, and knowledge.

We are now entering a period where human labour is becoming a rapidly diminishing component of the production process. Due to the exploitation of worldwide material resources they are equally available at almost the same price to any manufacturer. Thus competitive advantage rests almost wholly on the availability and efficient use of the financial and knowledge resources. Although this is an over-simplified analysis it would suggest that the emerging trends will place an increasing importance upon research and development (R&D).

Within R&D the microprocessor is already having an impact upon productivity. The time of the technologist is divided between the intellectual and creative aspects of the job and time-consuming computations and repetitive testing. The computational aspects have already been drastically simplified; now much of the routine work can also be automated. Simulations and computer-aided design also make a major contribution. Thus there will be a greater proportion of his time available for the intellectual/creative activities. This raises a number of questions about which one can speculate: How will this affect the number of research workers? – the demand for graduates and their calibre? Does the individual need periods of routine work interspersed with intense intellectual activity?

8. PERSONNEL AND MANAGEMENT

A continuing theme throughout this book and all that is written on microelectronics is the effect upon employment. Whatever one's views regarding the national prospects for employment there can be no escaping the conclusion that the individual company will maintain the trend towards lower employment in its manufacturing activities. This is now extending to the office and we have seen that in marketing, R&D, and elsewhere in the organisation there will be significant savings in labour.

Thus the number of employees in the firm will fall and so will all those activities directly associated with them – canteens, training, pay, and so on.

It also appears likely that the skills will polarise between the highly skilled and the unskilled. The effects of this upon the social organisation of the firm could be considerable.

The critical problem will probably be the recruitment and training of the highly skilled employees essential for the continued functioning of the operations. Boredom will not be removed, for much of their time will be spent waiting for something to go wrong. When it does they need to exercise their skills to the full. Fortunately this is not a new situation and the problems this type of work pattern presents can be overcome by studying the experience of process industries where a similar situation has existed for many years. This will, however, be a new situation for most manufacturing companies and could cause serious problems when the appropriate changes in organisation and managerial style are not introduced.

The role of the manager will be changed in many ways. In production he will be less concerned with the technicalities of the system and more involved with the behavioural aspects of his role (Twiss and Weinshall, 1980). For the key to success will be with the continued co-operation of a small but highly educated work-force.

In other functions much of the routine tasks will be removed from the manager. Thus we can expect to see a smaller number of managers partly because of the removal of routine, and partly because of the reduction in the number of the managed; but the emphasis will be on quality. We have already noted the premium on behavioural skills. In addition the manager will now be expected to use more sophisticated decision-making tools. In short the day of the professional manager will have arrived.

During the past 30 years a number of sophisticated managerial tools have been developed, particularly in the field of management science. Their application has, however, been limited. A major reason for this has been that they have not been suitable for use by the general manager and have become the province of specialists (e.g. OR, systems analysts). The cheap and convenient microcomputers and hand-held programmable calculators with their associated programmes will have a profound effect upon all levels of manager, who will now have easy access to the data he requires and the ability to process it himself in the way he needs for his decisions. But is this another false dawn? Pioneering efforts by a few large companies to train their staff in the use of these devices indicate that this is not the case, and that skill in using these aids will be expected of all managers.

CONCLUSION

The title 'The Managerial Dilemma' was chosen for this chapter because the advent of microelectronics raises significant questions to be answered by all levels of management and in all industries. Some of the effects of the new technology are already evident but there are many others about which we can only speculate. It is even now difficult to assess what the appropriate managerial responses ought to be.

In this chapter a number of these issues have been raised, many of which will be discussed in greater detail in subsequent chapters. Some indications of possible managerial implications and responses have been touched upon. But it would be dangerous to assume that there are simple answers of universal application. Every industry, every company, and each managerial situation is different and must be analysed accordingly. The only generalisation that can be made is that every reader of this book will be affected, more likely than not in the near future, and his response could have a significant influence on the future of his organisation. Because the impact is likely to be significant it needs to be thought through carefully – that requires time. Because of the speed of advances in microelectronics the time is short. This matter is urgent.

The important aspects of the introduction of microelectronics for management are not technical. The social impacts have been stressed. It is the impact on the total organisation which needs to be carefully planned. Although it might be wise to introduce the new systems in a limited way so as to gain experience where the consequences are not critical, it would be short-sighted to plan their long-term introduction without considering the effect on the total organisation. For in many instances a profound change in organisational structure and managerial style may be essential.

REFERENCES

Advisory Council for Applied Research and Development (ACARD): The Cabinet Office, *Computer Aided Design and Manufacture*. (HMSO, 1980).

Hedley, B. 'A fundamental approach to strategy development'. *Long Range Planning*, vol. 9, no. 6, December 1976.

Twiss, B. C. *Managing Technological Innovation*, 2nd ed. (Longman, 1980).

Twiss, B. C. and Weinshall, T. D. *Managing Industrial Organizations* (Pitman, 1980).

2 Electronics, Microelectronics and Microprocessors: What They Are

DENIS C. DALTON

1. INTRODUCTION

The electronics industry barely existed at the beginning of the twentieth century. Elementary telephones and gramophones were known but were in the main based on mechanical ingenuity. The existence of radio waves and the 'wireless' was known only to very few, and then with a degree of scepticism. Today every home takes some sort of electronic device for granted and knowingly or unknowingly every person relies upon electronics for the manufacture of everyday goods or the provision of everyday services. The international airways system could not operate, and billing for electricity, gas, and telephone could not be done without electronics. Indeed the majority of those in full-time employment would not be paid at all without the help of a computer. As a result of this technical progress a great worldwide industry has developed employing large numbers of people and creating vast wealth. There is no reason to suppose that the pace of technical development of the past 80 years will slow down. While everybody is already touched lightly by electronics, the developments that are 'on the drawing board' right now will influence everybody's lives even more profoundly.

This is technology, it is inanimate and has no 'common sense'. Whether its influence is for good or evil depends upon the use ordinary people make of it. Neither its development nor its influence can be stopped and consequently an understanding of it by all non-technical people is essential if they are to guide it in the right direction. The purpose of this section of this book is to lift a corner on this great industry and give some idea of what electronics and particularly microprocessor chips are all about.

2. BACKGROUND INFORMATION

2.1. ELECTRONS AND THINGS

In the beginning it all depends on electrons, so first we must understand a little bit about matter itself. A material like copper or a gas like oxygen consists of many small particles. In the case of a solid these particles are stuck together while in the case of a gas they are free and buzzing about at high speed. If such a material is divided and divided again into even smaller pieces there comes a time when it can be divided no more without losing its characteristics. This smallest particle into which matter can be divided without losing its identity is called an *atom*. Some materials consist of only one sort of atom and these are the '*elements*': copper, sodium and chlorine are examples. Other materials consist of combinations of atoms of such elements very strongly bonded together. Common salt is an example of such a *compound* and consists of sodium and chlorine combined in the ratio of one atom of sodium to one atom of chlorine. Salt has none of the characteristics of sodium or chlorine but a completely different set of its own. Indeed both the elements in salt are poisonous, yet salt itself is commonly used in cooking! The smallest particle of salt consists of one atom of sodium and one of chlorine stuck together. This particle is called a *molecule* and further subdivision will result in an atom of each of the constituent elements, and the disappearance of the salt.

The atom itself consists of discrete particles and is best thought of in terms of the solar system, which consists of the sun at the centre round which the planets (Earth, Jupiter, Saturn, and so on) revolve in a path known as an orbit, which may be circular or elliptical in shape. In an atom, the sun of the solar system is replaced by a very dense *nucleus*. One or more electrons replace the planets of the solar system – indeed they are sometimes called *planetary electrons* (Figure 2.1). A further piece of information needed to understand electronics is the concept of an electric *charge*. At some time or another everybody has combed his hair in dry weather and seen the comb attract the hair and make it stand up. The friction of passing the comb through the hair has left the comb holding an electric charge, and the hair with an equal and opposite

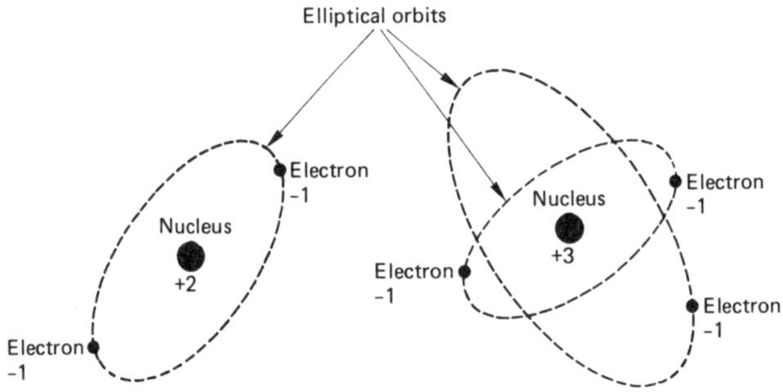

FIGURE 2.1 Helium has two electrons – lithium has three

charge. Any two bodies holding equal and opposite charge are attracted to each other. There seems a logic in nature that seeks to combine equal and opposite things to neutralise them and restore complete peace!

The nucleus of the atom holds a positive charge and the orbital electrons a negative one. Consequently they are attracted to each other and are only prevented from coming together by the centripetal force due to the rotation of the electron in its orbit. It is just like whirling a weight on a string round the head. The string is pulling the weight inwards and the centripetal force is trying to make it fly outwards. Again by the logic of nature the two forces are equal and opposite.

The difference between the atoms of different elements lies in the number of electrons in orbit round the nucleus. The simplest has only one and others have many. The charge on one electron is always the same and the atom as a whole is never charged. Consequently the charge on the nucleus is always equal to the number of electrons in orbit.

2.2. CONDUCTORS AND INSULATORS

Materials having many electrons in orbit do not have them in the same orbit. Some electrons follow a path relatively near to the nucleus while others are relatively far away. Those in the outer orbit may sometimes be attracted away by another oppositely charged atom nearby. Such a free electron thus may join a nearby atom which is short of an electron, neutralising the positive charge on it, and leaving a positive charge on the atom it has just left. This action can have a chain-like effect throughout the multitude of atoms forming the material, and charge can flow from one end to the other. Such a material is a conductor of electricity. The most common such conductor is copper, universally used in the wire or flex of domestic equipment, but there are many other conductors – some like silver and gold are better than copper while others like iron are poorer. All, however, have the common characteristic of so-called free electrons in the outer orbits of their atoms.

Other materials, however, have no such readily detached electrons and are therefore incapable of conducting electricity. These are used as insulators, to prevent electricity flowing. Again in the domestic equipment flex the insulating covers were originally rubber and cotton, but today man-made materials like polythene, PVC or nylon are used.

There is a third category of materials which, while able to conduct electricity, are a significantly poorer conductor than the metals but much better than polythene and PVC. These are known as semiconductors, the most important of which are the elements silicon and germanium.

2.3. THE VALVE OR VACUUM TUBE

Scientists who at the turn of the century had invented the telegraph or telephone, both of which depend for their operation upon the conduction of electricity through copper wire, did not understand that the atom consisted of a nucleus and orbiting electrons. Consequently they believed that electricity flow happened because of the movement of charged atoms. It therefore followed that a vacuum could not be a conductor, since it contained nothing. As soon as charged atoms were introduced it would cease being a vacuum. Imagine the surprise when, quite by accident, flow of electricity was observed by experimenters working with evacuated electric light bulbs. Furthermore it was found that the ability of such a vacuum to conduct could be changed by introducing an electric field (a term used to describe the imaginary pieces of elastic which draw two oppositely charged bodies together). This observation led to the development of the first thermionic vacuum tube. This consists of an evacuated glass envelope containing a filament similar to that in the light bulb, and a plate connected externally to the filament through a battery which maintains it positively charged relative to the filament. The electrons which bubble off the filament due to the heat are immediately attracted to the plate and current flows as shown by the arrows (Figure 2.2). (Conventionally current, being the flow of negatively charged electrons, is in the opposite direction to the electron flow.) If now a mesh is introduced between the filament and the plate, the movement of the electrons and hence the current flow may be controlled

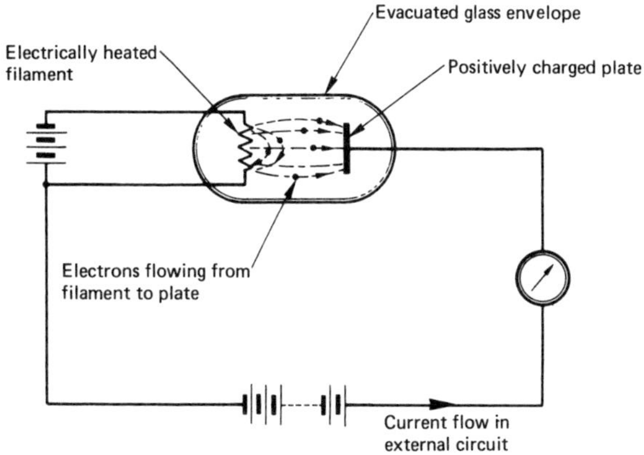

FIGURE 2.2 A thermionic diode

by changing the charge on this mesh or grid. This fundamental phenomenon is the origin of all electronics, as it can be used to amplify a signal and to switch and control currents in a way never before possible.

The electronics industry grew up around sophisticated versions of the elementary thermionic tube just described and communications and control in World Wars I and II were based on it. The tube suffered three great disadvantages:

(a) it was very large;
(b) it was very expensive;
(c) it consumed much power and got very hot.

These disadvantages prevented further progress and the electronics technology had almost reached its limit. Then in 1948 Bell Telephone Laboratories demonstrated the transistor.

2.4 SEMICONDUCTOR DEVICES

2.4.1 P- and N-type semiconductor materials

The transistor is made from semiconductor materials already mentioned. Almost universally silicon and germanium are used. The transistor's usefulness is exactly the same as that of a vaccum tube – it can control a relatively large current by varying a relatively small voltage and can thus amplify an input signal.

We have already seen that the atoms of a solid material are firmly stuck together. To understand a transistor it is necessary to know why.

An atom consists of a nucleus and a number of planetary electrons orbiting round it. The electrons, however, follow a predetermined path or orbit which are of a predetermined size and shape. Each orbit can contain no more than a given number of electrons: for example the inner orbit can contain no more than 2 electrons. Silicon has fourteen electrons; two in its inner orbit, eight in a middle orbit, and 4 in the outer orbit. It is the four in the outer orbit that are of interest to us, because these outer, or valence, electrons determine the majority of the element's electrical and chemical properties. When the atoms of an element are forced close together, as they are in the case of a solid, they frequently align themselves in a uniform three-dimensional pattern and form a crystal or lattice structure. The four outer electrons of the silicon atom form pairs with one of each of its four neighbours. This pairing of electrons is what is called the formation of covalent bonds, and is why the lattice is rigid and the element is a solid. This formation of bonds is illustrated diagrammatically in Figure 2.3. A piece of silicon large enough to be seen would consist of untold millions of these atoms bound together in this way.

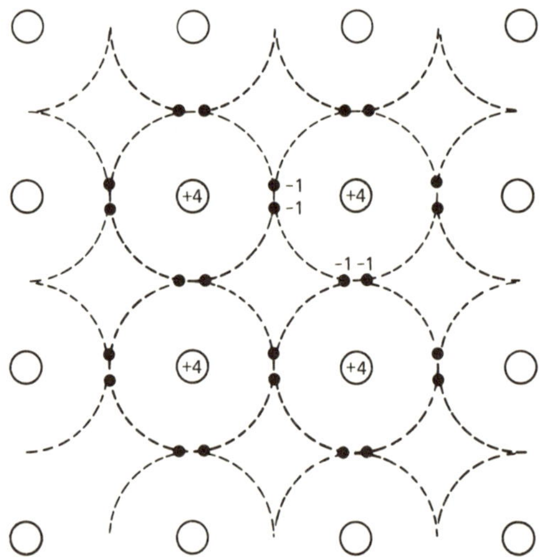

FIGURE 2.3 Covalent bonds

If, however, one of the electrons forming a covalent bond acquires more energy, from the heat of its surroundings or some other source, it can break away from its bond and wander through the structure in a more or less aimless manner. In the same way the bond from which the electron has escaped is left deficient of one electron. This deficiency is termed a hole, and it too can move through the structure. The hole is effectively a moving positive charge.

A more efficient method of producing electrons in a silicon crystal than by using heat and light, is by the use of donor impurities. A very small amount of phosphorus, which has five valence electrons in its outer orbit, will enter the structure, form the necessary four covalent bonds with its neighbouring silicon atoms and have one free electron left over. This is only loosely bound to its nucleus and is easily made available to move about. Silicon containing phosphorus is known as N-type silicon, referring to the fact that conduction is done by moving electrons which of course hold a negative charge.

Impurities like boron, gallium, and indium have only three electrons in their outer orbit and if such an atom replaces a silicon one in the structure, it can form only three covalent bonds. There is therefore a deficiency of one electron or a hole. Silicon containing such hole-producing impurities is known as P-type, referring to the fact that conduction is done by moving holes which hold a positive charge.

2.4.2. The PN junction diode

If now we take a section of N-type and a section of P-type silicon and join the two together we obtain a device with peculiar characteristics. When these sections are first brought together each is electrically neutral. The free electrons in the N-type seek to fill the holes in the P-type until the N-type loses some electrons and becomes positively charged, thus restraining further electrons from crossing the junction. In similar fashion loss of holes in the P-type leaves it negatively charged, thus further preventing electrons crossing the boundary. The region in the immediate vicinity of the junction has thus lost all its mobile charge carriers and is called the depletion region. Current can only be made to flow (i.e. electrons caused to cross the junction) by overcoming the positive charge on the N-type and the negative charge on the P-type by means of an external source of potential difference (a battery will do). The current flow in such a forward-biased junction increases quite sharply as shown in Figure 2.4. If, however, the battery is reversed the external potential difference is aiding that produced internally and, to a first order anyway, no current will flow. A P–N junction diode is therefore a one-way device, having an overall characteristic as shown in Figure 2.5. The small current which flows in the reverse direction is due to secondary effects in the material which need not concern us here.

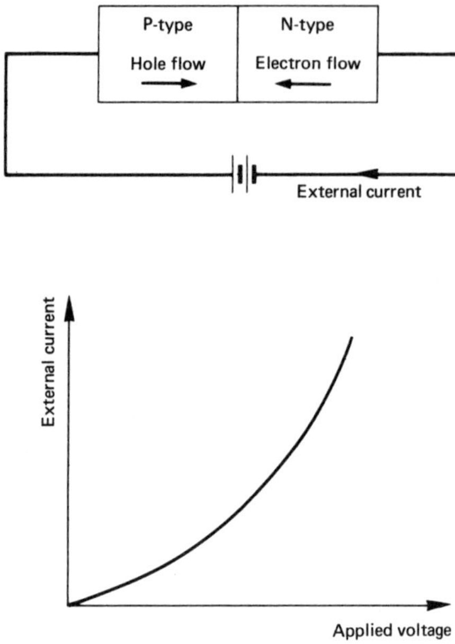

FIGURE 2.4 Electrons and holes in a semiconductor diode

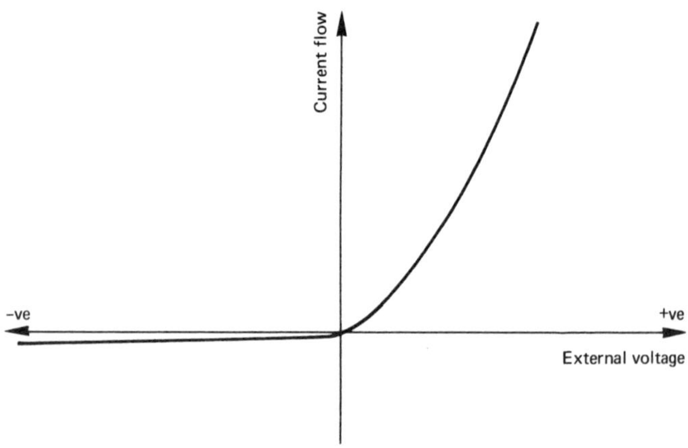

FIGURE 2.5 Characteristics of a P–N junction diode

If now an alternating external voltage is applied to such a P–N junction diode, current will flow during that part of the cycle which biases the diode in the forward direction but very little will flow when the applied voltage reverses during the second half of the cycle (see Figure 2.6). This diode effect produces a one-way device and is very useful in electronic circuits.

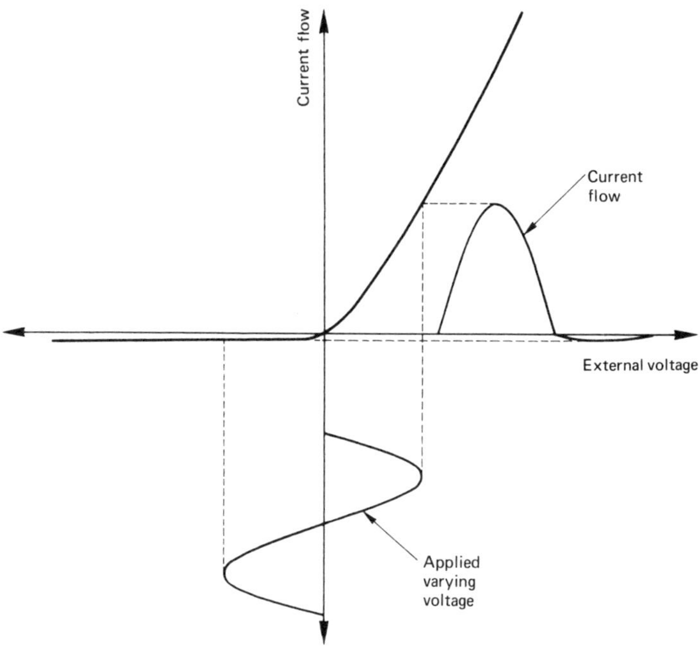

FIGURE 2.6 Diode effect in a P–N junction

2.4.3. The N – P – N transistor

If now a further section of N-type semiconductor is added, thus making a sandwich of the P-type section, depletion regions are formed at both junctions between N- and P-type material. If an external voltage is applied in the manner shown in Figure 2.7, one of the junctions will be biased in the forward direction and current will flow, whereas the other is biased in the reverse direction and no current will flow. If, however, the P-type sandwich is made very thin a further unexpected phenomenon occurs. Electrons leaving the N–P junction which has a bias in the

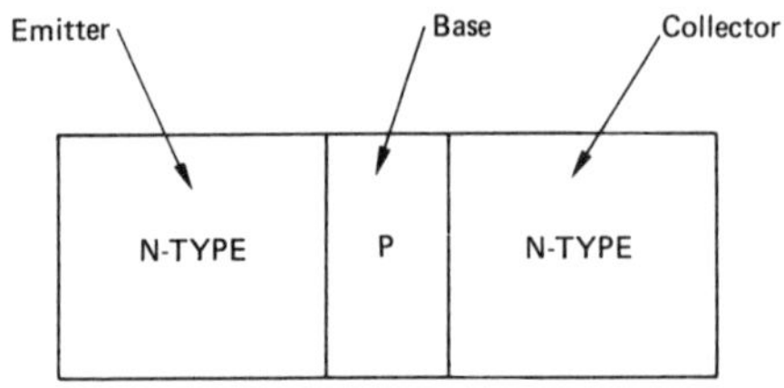

FIGURE 2.7 N–P–N transistor

forward direction travel fast enough to pass right through the P-type region as a result of their own inertia and despite the reverse bias across the other junction. Once they arrive in the other section of N-type material they find a positive field from the biasing battery which helps them on their way.

Whether or not these electrons succeed in passing right through the thin sandwich of P-type material depends on the bias upon it. Hence the current which passes right through the device can be controlled by the voltage on the P-type material. Each of the sections of such an N–P–N transistor have been given names. The sandwich of P-type material is called the base, the N-type sections are respectively called the emitter and collector. These terms are self-explanatory if it is remembered that they refer to electrons. The flow of electrons is from emitter to collector. The flow of current, as conventionally understood, is in the reverse direction.

This transistor effect can produce the effect of amplification and is the basis of all modern microelectronics and computer technology. Without it all modern information, collection, transmission, and processing systems would have been impossible.

2.4.4. Microelectronics

We have seen how a valve used for amplification consisted of wires mounted inside an evacuated glass envelope. It was intrinsically a big, power-hungry device. The transistor, on the other hand, consists only of a very small piece of silicon. By changing the geometry of the early transistors and by using modern engineering techniques, it is possible to take a thin slice of pure silicon and diffuse the impurities required to produce the N-type regions and the P-type regions into the surface. Furthermore by the use of precise photographic and lithographic techniques these can be made very small indeed, to the point that is now not uncommon to find 5000 to 10,000 such transistors constructed on a piece of silicon 5 mm square and a fraction of a millimetre thick! It is by organising this sort of complexity that modern solid-state microelectronic devices are formed. Each device on the chip is a simple transistor similar to that already described, coupled together by means of interconnecting links of material of high or low conductivity. Nevertheless despite the simplicity of the basic transistor, the complexity available has made possible fast and accurate handling of information.

3. COMPUTERS

Much has been written about the microprocessor. In construction, it is a particular form of microelectronics. To understand why it is of such great significance we must know a little about the basics of computer technology.

3.1. PROCESSES AND PROCESSORS

What is a process? One dictionary definition is;

> A series of changes or measures.

The word has been adopted by engineers in industry, and in so doing its meaning has become rather more specific. Thus it could be defined as

> A series of actions or operations leading to some expected result.

The shift of emphasis on to the idea of doing something definite with the expectation of a known result is a key. A tangible object could be called a processor if it

> *Carries out a series of actions or operations which lead to a known or expected result.*

A particular process yields a particular result but a given result may be obtained by one of several different processes. The familiar phrase which says the same thing is 'There are more ways of killing a cat than drowning it.' The process is drowning, the expected result is to kill the cat. Changing the process to shooting it would do equally well and produce the same result.

Processors occur in profusion in nature. Some are so small they might even be called 'microprocessors'. Two examples are:

(a) The green colour of a leaf of a plant is due to the presence of chlorophyll. In the sunlight, the leaf collects carbon dioxide from the atmosphere and the chlorophyll processes it so that carbon compounds needed by the plant are synthesised. A by-product is the unwanted oxygen, which is released back into the atmosphere thus accidentally enriching its oxygen content to the benefit of oxygen-breathing organisms. This process is called 'photosynthesis'.
(b) The yeast which occurs naturally in the bloom on fruit cannot survive by breathing oxygen. It obtains it by processing the sugar in the fruit. By-products of this process are alcohol and carbon dioxide. The result is survival for the yeast but decomposition of the fruit. This process is called fermentation.

Man has, over the centuries, been very astute in turning processors occurring in nature to his own use. Yeast, for example, is used in baking bread. The by-product of fermentation is carbon dioxide gas which cannot escape from bread dough and forms little bubbles within the material. When the dough is baked, the yeast is killed by the high temperatures but the little holes remain in the solidified dough giving it its light and pleasant texture. Yeast's other by-product is also widely used by man. The alcohol produced in fermenting grape juice eventually reaches such a strength that it kills off the yeast and a pleasant drink is produced. If the fermentation happens in a sealed vessel the carbon dioxide cannot escape, is forced to dissolve in the wine and reappears as bubbles when the vessel is opened. Champagne is the result of this process if appropriate ingredients and patience are utilised.

Notice, however, that both the processors discussed can only carry out one process. Chorophyll can only photosynthesise, and yeast can only ferment. They are therefore dedicated to one purpose and are called 'dedicated' processors.

Undedicated or programmable processors are a more recent phenomenon. It takes a very high intelligence to be able to create options and to make a choice between them. A typical man-made processor is the water wheel, whose job in its irrigation role is to scoop water out of a pond or stream at a lower level and dispose of it into an irrigation channel at a higher one. In this simplest form the water wheel is dedicated to the single task of raising water into one channel. With a little ingeniuty, however, the chutes in the water wheel may be made to tip from side to side and to discharge their water into one of several such irrigation channels. Such rocking chutes would tip in one direction or another quite at random unless a cam or other device is provided to turn them in a chosen direction. Such a cam could make one chute turn one way and the next two the other, so that one channel receives half as much water as another. Changing the cam to one of a different shape could change the proportions of water delivered to the different channels. Such a water wheel is still a processor, but it is programmable or undedicated – changing the cam changes the programme. Man has designed in some options and a means of exercising choice. Now we can define the difference between a dedicated and undedicated processor thus:

A dedicated (non-programmable) processor carries out only one series of actions or operations

An undedicated (or programmable) processor carries out actions or operations as instructed by a programme – which may be changed.

3.2. THE IMPACT OF ELECTRONICS

Man has always been very ingenious. Using the elementary technology at his disposal he has been able to design some very engaging processors, but due to the size of the various parts and the difficulty of making them most early processors were dedicated. They focused on processes not necessarily concerned with material things, but more with ideas and numbers. Typical is the early Chinese abacus used as an aid to mathematics and the sand-timer (egg-timer) used as an aid to the measurement of time. A more recent example of man-made dedicated processor is the device used to govern the speed of a steam engine (Figure 2.8). As the engine speed rises the weights tend to fly outwards due to centripetal force. This causes the disc to rise and compress the spring. The valve controlling the amount of steam fed to the engine is closed as the disc rises, and in turn therefore the engine slows down, the weights move inwards and the disc down, and the steam valve opens. Obviously a position of equilibrium is reached and the engine runs at this speed constantly. The governer is processing information (i.e. engine speed) and producing a result (opening or closing the steam valve to keep engine speed constant).

When the thermionic tube was invented, it was realised that an electrical analogue representing engine speed could be manipulated by electronic circuits and a signal sent to an electrically operated steam valve to open and close it appropriately. In a similar way an electronic equivalent of an abacus can readily be conceived, making use of electronic vacuum tubes. In both cases a practical electronic equivalent

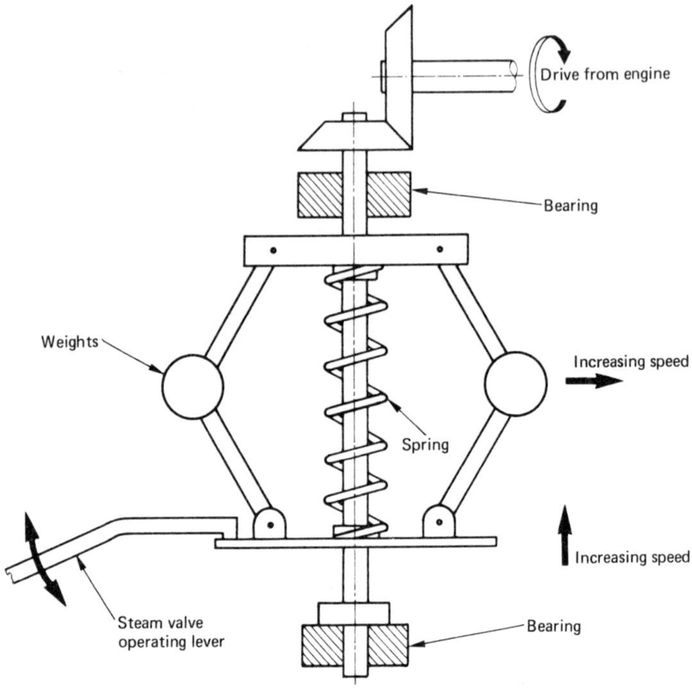

FIGURE 2.8 Steam engine speed governor

was large, expensive, unreliable and consumed much power. Laboratory samples of such electronic control and calculating machines were made between the wars but their cost and size prevented them from being used in everyday applications.

We have already seen that the invention of the transistor and its successor the integrated circuit, having many thousands of transistors on a single small chip of silicon, overcame the disadvantages of the valve while retaining its potential to manipulate information electronically.

3.3 ANALOGUE OR DIGITAL?

As has already been mentioned, the manipulation of information by electronics depends upon generating some electrical representation of the quantity or fact. In the early days of wireless, for example, useful information was sent from the transmitter simply by turning it on and off according to some predetermined code and by using the receiver to detect whether the transmitter was 'on' or 'off'. The code most commonly used, and which was standardised internationally, was the Morse code. This consisted of combinations of short (dots) and long (dashes) bursts of transmission. Thus the morse code representation of the alphabet is given in Figure 2.9. As radio technology developed it was found possible to do something more flexible than to turn the transmitter on and off. It became possible to vary the strength of the transmission at a controlled speed. It was thus possible to send two pieces of information at once: pitch (frequency) and loudness. Thus the very complex shape of the vibration produced by the voice or by musical instruments could be converted into an electrical analogue, transmitted, received, and converted back into audio vibrations. Early information processors (now called computers) relied upon the use of such electrical analogues of information and were known as analogue computers.

A	·—	N	—·
B	—···	O	———
C	—·—·	P	·——·
D	—··	Q	——·—
E	·	R	·—·
F	··—·	S	···
G	——·	T	—
H	····	U	··—
I	··	V	···—
J	·———	W	·——
K	—·—	X	—··—
L	·—··	Y	—·——
M	——	Z	——··

FIGURE 2.9 The Morse code

The modern computer works on a code similar to the Morse code except that only dots are used. The basic reason is that it is very easy to count with electrical circuits that are either 'on' or 'off', whereas it is difficult to determine whether a circuit is six-tenths or five-tenths on! Consequently an arithmetical convention has developed of representing numbers as a series of noughts and ones, nought being circuit 'off' and one being circuit 'on'. Such a nought or one – known as a binary digit or 'bit' – can be used for counting in exactly the same way as the more familiar arabic numerals normally used. The numbers 0 to 9 and their binary equivalents are given in Figure 2.10. By using the normal rules of arithmetic binary numbers can be added, subtracted, multiplied and divided in exactly the same way as our familiar decimal numbers. To the human being they seem excessively long and complicated to deal with, but to an electronic computer they are easier.

0	0000
1	0001
2	0010
3	0011
4	0100
5	0101
6	0110
7	0111
8	1000
9	1001

FIGURE 2.10 Binary notation

As the Morse code allocated meanings to intrinsically meaningless sequences of dots and dashes, so enabling the letters of the alphabet to be transmitted, so it is possible to allocate meanings (letters, words, and any other symbols like ×, +, etc) to sequences of bits. Information coded in this way is known as digital information or data, and a computer designed to process digital information is referred to as a digital computer. It is safe to say that all modern computers, big or little, are digital.

3.4. THE COMPUTER AND THE MICROCOMPUTER

Now a definition of a computer can be written which uses words which have already been discussed.

A computer is an electronic machine which can carry out a specified programme of simple arithmetical or logical instructions without operator intervention.

A computer consists of three essential parts (Figure 2.11):

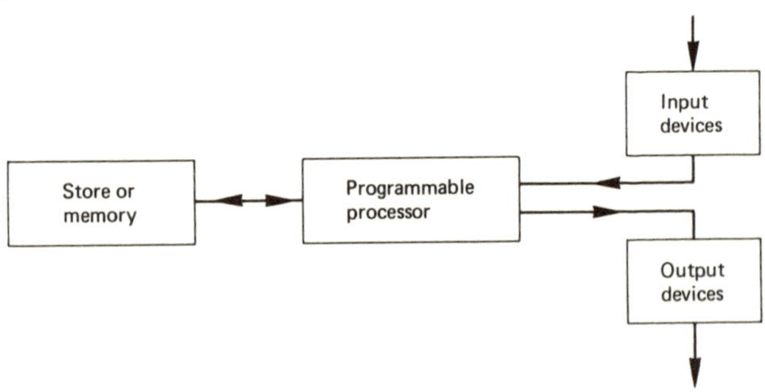

FIGURE 2.11 A computer's essential parts

(a) a programmable processor (or undedicated processor);
(b) a store or memory (to store program and input/output data)
(c) input and output devices (for communicating with the outside and non-electronic world).

Each of these parts will be discussed more fully later. Both the processor and the memory can, using recent silicon technology, be made to be very small and compact. A microprocessor is merely

A tiny electronic digital programmable processor fabricated on a single chip of silicon.

When a microprocessor has memory and input/output devices associated with it, it becomes a microcomputer, having much the same capabilities as the older computers.

It can be seen now that there is very little new about the microprocessor, despite all the sudden interest taken in it by the lay press. It is part of the continuum of electronics which started with the spark radio transmitter at the turn of the century; but the progress in present-day electronics technology has caused size and cost per function to fall and reliability, complexity, and therefore capability to rise so that the point has been reached that electronics generally can be built into commonplace articles providing increased capability at little or no extra cost. It is at such times that traditional products and markets undergo sudden and radical change. Already both watch and calculator industries have been completely transformed. The telecommunications industry is undergoing revolutionary change at the present time and electronics is becoming so pervasive that it is exceedingly difficult to predict its impact on other important industrial activity.

So how does a computer, and for that matter a microcomputer, operate? In essence it is very simple. Its operating sequence or machine cycle consists of only three steps, namely:

(a) obtain the next instruction from memory;
(b) read and decode it;
(c) implement it.

Nevertheless a computer is very complex; the reason is its scale and speed. A machine cycle takes only a few microseconds to perform (a microsecond is one millionth of a second). Indeed computers of the

1980s may well perform 100 million cycles per second. A program of considerable length (say 5000–10,000 steps) may still take only a fraction of a second to perform, and repetitive methods unacceptable to the human brain are easily and quickly handled. Equally if so many machine cycles are to occur in a second a great deal of information can be absorbed, operated upon, and stored ready for use. This has given rise to advances in the technology of memory to keep up with the progress of processors. It is therefore appropriate to look at microprocessors, memory and input/output devices rather more carefully.

3.5. THE PARTS OF A MICROPROCESSOR

The microprocessor consists of a number of different parts, each with their own job to do, but all contained on the single chip of silicon. Typically a microprocessor will contain the following sections:

(a) *Accumulator register*
This is the main work register into which data and results are written and processed, and from which they are subsequently despatched to memory or output ports.

(b) *Program counter*
The instructions which form a program must be retrieved from the program storage memory in sequence if the operation is to be carried out correctly. The main function of the programme counter is to keep a count of where it is in the programme.

(c) *Instruction register*
When an instruction is retrieved from memory, it is stored in the instruction register which holds it for decoding. The simple processor using an 8–bit instruction code would have a capacity of up to 256 separate instructions.

(d) *Instruction decoder*
This decodes the instruction and tells the central processing unit (CPU) what to do.

(e) *Scratch pad memory*
This is used for all sorts of temporary storage and is usually organised so that access is easier than to the main memory area. It will contain commonly used addresses and frequently is so sequenced that successive memory locations can be addressed by simple incrementation.

(f) *Arithmetic/logic unit*
This is the section which actually performs the computation.

(g) *Clock and control circuitry*
The clock generates the time operation for the whole processor system and is usually determined by the speed at which the various parts of the processor can function. The actual sequence of events in the processor is organised by the control circuits.

Most microprocessor chips available at present consist only of the processing unit and must have associated with them memory and input/output devices to become useful. Some more advanced chips already contain simple memory and input/output electronics and this trend will continue in the future.

3.6. MEMORY

In the past, computer memories frequently consisted of bulky ferromagnetic devices. Modern technology has enabled those memories closely associated with the processor to be made on semiconductor chips.

Although there are many ways of realising memory in semiconductors all current technology falls into two basic classes:

(a) *Read only memory (ROM)*
This is capable of storing a series of bits forming instructions or information in accordance with predetermined data, but once coded cannot easily be altered. Indeed, some cannot be altered at all.

(b) *Read–write memory*
This is also capable of storing series of bits of information but the data stored can be changed by a microprocessor within the time available in a machine cycle. Access to a read–write memory is nearly always required to be in some random order dictated by the programme. They have therefore become known as random access memories (RAM) despite the fact that access to some ROMs must also be random.

To operate a microprocessor, two sets of data need to be stored:

(a) *Programme memory*
This would consist of a basic instruction set and the programme instruction set which is inserted and related to the special operation under consideration. The first of these is nearly always in ROM, the second in RAM if the microcomputer is undedicated but can be in ROM if it is dedicated.

(b) *Data memory*
This has three jobs: the storing of input data, the temporary storage of interim data in the course of the programme, and storing the processed data which is to form the output. All this data must be changed by the microprocessor during the course of its operations and therefore data memory must be in RAM.

3.7. INPUT DEVICES

Inputs to a microcomputer, apart from those already stored in ROM and inserted by the manufacturer, may be from a variety of sources:

(a) Secondary memory, e.g. magnetic disc, magnetic tape, punched card, punched tape, new technologies (e.g. bubble memories, holographic stores).
(b) Another computer or microcomputer.
(c) Sensors, e.g. strain gauges, thermometers, electrical measuring devices.
(d) Keyboards.

Clearly, suitable conversion is needed to give the microprocessor a digital input that can interpret, and all input devices may be remote from the processor and connected to it by communication links such as those operated by the Post Office. The computers on board the moon spaceships were connected to the earthbound space headquarters by radio data links.

3.8. OUTPUT DEVICES

These include:
(a) alpha numeric displays using LEDs (light-emitting diodes) or liquid crystals, as seen in a digital watch, or hand-held calculator;
(b) character or line printers on paper;
(c) visual displays using cathode ray tubes;
(d) other secondary memories as described above;
(e) actuators, e.g. switches, valves, etc. with suitable interfaces.

Thus, despite much talk in the press about a miracle on a chip, it can be seen that a microprocessor is only part of an operating system needing to have other hardware associated with it if it is to do a useful job.

4. WHAT A MICROCOMPUTER WILL DO

A microcomputer will:

(a) TAKE A SET OF DATA, SUITABLY ENCODED, AND STORE IT

As already explained this data consists of bits of digital information. To a string of 0's and 1's significance can be given like the numbers shown in Figure 2.10, or letters used in building up words, or any other consistent meaning. To the microcomputer such strings of 0's or 1's have no meaning; the human being has assigned such meanings.

It can be seen therefore that the memory needed to store such data must contain one cell for every bit to be stored and that since such data uses long strings of 0's and 1's a very large number of such cells are needed. It is commonly possible to store 32,000 or 64,000 bits of information on one chip and the capability to store 1,000,000 bits on a chip is within the scope of the technology.

(b) RETRIEVE ANY PART OF THIS DATA AND PERFORM SIMPLE FUNCTIONS ON IT

A string of 0's and 1's of a given length, usually 4, 8, 16, or 32 bits depending on the microcomputer, is called a word of data. Each word stored in memory is stored in a location having an address. By merely specifying the address it is possible to retrieve the word. Then as an example the instruction may be to double its size and re-store it in the same address. There could be several hundred such instructions available but the human programmer is needed to specify which, and to arrange these in the logical sequence needed to give a desired result.

(c) DO THE SAME EVERY TIME

When called upon to carry out an instruction such as multiply by 2, human beings frequently 'get it wrong'. In a properly constructed microcomputer this never happens. So while each instruction is only simple one, a microcomputer can be relied upon to 'get it right' every time.

(d) PERFORM AT VERY HIGH SPEED

As already explained the machine cycle:

- obtain the next instruction from memory;
- read and decode it;
- implement it;

takes only a few millionths of a second to perform. A programme of 10,000 steps – so long that it is difficult for a human being to read and understand in one go – will take only a few hundredths of a second to carry out. A programme of this length is sufficient to calculate the logarithm of a number from first principles! It would be daunting to a human to have to calculate his logarithm every time he wanted to use it! The microcomputer therefore is very good at performing complex and repetitive operations.

(e) DEAL WITH A CHANGING SITUATION

Because of this speed, the output of a microcomputer changes almost immediately if the input changes. There is of course some delay but this is short and in most applications insignificant. If therefore an actuator is to be moved as a function of a changing input according to some complex function, the reaction to change is effectively immediate. An example of such a feedback system occurs in the motor car fuel system. By measuring the contents of the exhaust gases and adjusting the fuel injection system accordingly it is possible to maximise miles per gallon and minimise unwanted exhaust emissions. The link of logic and calculation between the exhaust sensors and the fuel injectors is admirably provided by a microcomputer. Suitably programmed microcomputers may therefore be placed into the feedback or control loop of a system and be relied upon to deal immediately with any situation the programmer has anticipated – *but no others.*

(f) GIVE THE IMPRESSION OF THOUGHT

Basically a microcomputer will do only what a programmer has previously enabled it to do by programming it. The technology of programming has not yet reached the stage of enabling a microcomputer to programme itself. This means that it cannot in anyway resemble 'thought' as humans know it, but sometimes its speed may give the impression of thought. This is especially true when the microcomputer is provided with a programme which causes it to do one thing if a certain set of conditions are met, and something totally different if they are not. Such conditional jumps, as they are called, may occur frequently in a complex programme. When such a programme is run it gives the impression that the microcomputer is intelligent; and therein lies the trap for the unwary. If a microcomputer is to perform a useful task, it must be programmed with a degree of precision rarely found in systems operated by people. The manager relies on the common sense of his staff to correct the flaws in his instructions; the engineer relies on approximations to avoid tedious and unnecessary precision. The microcomputer will do neither of these things. Consequently it will not reduce the cost or time of solving a problem, it will only remove the drudgery involved in solving similar problems many times over. The microcomputer is therefore useful in any application when information is to be manipulated quickly accurately, and according to predetermined rules.

5 THE FUTURE

Much has already been written about the social and economic consequences of the microprocessor, and later chapters of this book speculate further on these topics. But has the technology reached its limit?, are these speculations based on a fixed technological scenario or are they likely to be invalidated by further rapid advances in technology?

The facts are that microprocessor, memory, and other microelectronic technologies are advancing at a very high rate. The present technology is only a snapshot of a continuous advance from the early thermionic diode described earlier to an as yet unknown limit. So why all the recent excitement about microprocessors? It is suggested that the steady fall in size and cost per function, coupled with marked increases in reliability, complexity and therefore capability have led to the point where microcomputers are available economically for use in commonplace applications.

At present the performance of microcomputer-based products is limited mainly by the capability of the product designer and in particular of the programmer. As already explained the programmer must be both precise and knowledgeable of the application, as well as being familiar with computer technology. Such people are in very short supply and the number and scope of microcomputer applications are limited more by the shortage of this resource than by the capability of the technology.

The technology itself will continue to progress. Devices will become ever faster in operation and smaller in size. It has already been mentioned that 1 million bits stored on one chip is within sight. It is anticipated that the computer of the 1990s will:

(a) carry out 70 million instructions per second;
(b) store 64 million bits of information;
(c) occupy only a 6-inch cube of silicon.

Clearly the availability of such performance underlines the statement already made that applications will not be limited by the technology, but only by the wisdom, accuracy, and imagination of the people who seek to use it.

3 Strategic Implications of Microelectronics

PETER TRIER

INTRODUCTION

The words 'microprocessor' and 'microelectronics' are not two expressions for the same thing. Microelectronics is a large subject of which the microprocessor is an important embodiment. In fact, the microprocessor has become a powerful and emotive symbol of the technology of microelectronics. But although the microprocessor is only one embodiment, it is possibly the most important one in its range of applications, particularly applications outside the traditional realm of electronics, and also as a product of the marriage between semiconductor and information technology. It enables us to combine, in a single compact electrical circuit, activities which have traditionally been handled separately, i.e. the functions of Information, Communication and Control.

Information has evolved from the manuscript, the printed book, the catalogue, and from there to its electronic embodiment in the computer store. *Communication* has developed as the letter, the telegram, the telephone conversation, and the many modern outgrowths of communication from one person to another. *Control*: the reins on a horse are one of the earliest control mechanisms . Automatic control mechanisms follow, like the centrifugal governor on a steam engine, or the thermostat which controls the temperature of a car radiator.

It is only within the last few years that these three functions have come together in integrated electrical systems, and it is this integration that is the essence of the new industrial revolution. With muscle power having been replaced by machine power, today these three constituents make up most industrial work whether in manufacturing industry, energy supply, process industry, transport, service industries, office work, defence systems, or telecommunications themselves. Information, communication, control, dominate activity in all these fields. In these fields, electronic systems play an ever-increasing part, but they are also becoming important in medicine, in education, in the home, and not least, in the motor car.

In mentioning the new industrial revolution, one can nowadays differentiate not less than three industrial revolutions. The first one lasted from broadly 1800 to 1900, and was largely based on steam. The second one has been described as the electromechanical revolution, power becoming available in 1880, and electrical communication becoming important in 1900. Early electronics, embodied in the radio valve, led into control systems in the 1920s. Information processing started in the 1950s, and that led gradually into the third industrial revolution, the microelectronic systems revolution, in which information systems, control systems, and communication systems, are no longer separate but integrated. This third industrial revolution is based on three distinct elements: firstly, semiconductor technology – the silicon revolution; secondly, the concept of digital information processing – the information revolution; and thirdly satellite technology – the space revolution. Let us look at them in turn.

THE SILICON REVOLUTION

The silicon revolution started with the invention of the transistor; a solid-state amplifier, small, compact, to take the place of the old-fashioned radio valve. Then came the integrated circuit, which enabled a number of transistors to be built on a common substrate by simultaneous manufacturing processes, based on photo-lithographic techniques. Making integrated circuits is akin to the process of multicolour printing, although with very fine dimensions.

The third step, the 'large-scale integrated circuit', came in 1970, when we got from tens of circuit elements per silicon chip to thousands; and so into the era of the VLSI, the 'very large-scale integrated circuit', where

TABLE 3.1 *The silicon revolution*

The transistor	1948
The integrated circuit (IC)	1960
The large-scale IC (LSI)	1970
The very large-scale IC (VLSI)	1978

we make tens of thousands, hundreds of thousands, and before very long millions of circuit elements on a small, single silicon chip. The increase in density, which goes with this increase in the number of elements, is also accompanied by an enormous decrease in price per element, per logic gate, or per bit of memory storage.

The density of memory storage is illustrated by simple statistics. In 1965 it took 11,000 litres of volume per million bits of computer storage, including associated circuits. In 1977, that had been reduced to about 1 litre per million bits. But, to stop us getting complacent, the human brain is still ten thousand times better. The brain stores a million bits of information in one ten-thousandth of a litre.

The price-lowering in the face of the expensive technology needed for all these achievements can only be obtained by an explosive expansion of demand, of manufacturing capacity and rationalisation.

Here is one important documentation of the increase in demand for building element functions from 1960 to 1985, in 5-year steps. 'Building element functions' are, for example, a single gate or a single storage element or a flip-flop, i.e. any elementary function (Table 3.2).

TABLE 3.2 *Expansion of electronics: world use of building element functions*

1960	3×10^8
1965	3×10^9
1970	3×10^{10}
1975	3×10^{11}
1980	3×10^{12}
1985	at least 3×10^{13}

i.e.: × 10 every 5 years; × 100 every 10 years; × 1.6 from year to next year.

This table provides impressive statistics. Of course, the total value of the semiconductors used increases much more slowly, because of the price-fall mentioned; but as these components only make up a part of the equipment or the complete system, the total delivered value of an electronic system increases quite a lot in spite of this price-fall. Table 3.3 shows an estimate of the world electronics market in 1966, 1976, and 1986. The expansion from 1966 to 1986 is impressive in every category; in communications and computers it is explosive. Expansion of the total world electronics market from 1966 to 1986 is going up in spite of all the price-falls, from 48 billion dollars to 480 billion dollars, a factor of 10. 'Volume' expansion is difficult to define accurately with changing

TABLE 3.3 *Estimated world electronics market* ($\$ \times 10^9$)

	1966	*1976*	*1986*
Telecommunications	17	52	160
Data processing	5	36	150
Consumer goods	11	28	72
Instrumentation	5	16	50
Military	10	22	48
TOTAL	48	154	480
COMPONENT CONTENT	12%	10%	8%

technologies, but will certainly be even larger, so that when people talk about gloomy employment prospects, always bear in mind that electronics will go on in a big expansion for many years to come. It is interesting that the contribution of components in all this market is slowly on the decrease, because components are getting cheaper so rapidly.

This is illustrated in Table 3.4, showing the price fall of memory storage from 1970 onwards. Prices before 1970 are not shown, but the downward trend was already rapid, and not very long before, the cost of storage per bit was $2 or more.

TABLE 3.4 *LSI memory circuits*

	Chip capacity (bits)	*Chip size (mm^2)*	*Cost per bit (US $)*
1970	1,000	3	2
1976	16,000	20	0.2
1980	250,000	30	0.02
1983	1,000,000	40	0.01

Yet another example of cost reduction, arising from technical innovation, but taken from the user field, is the transatlantic telephone call (see Table 3.5).

THE INFORMATION REVOLUTION

I have already mentioned the digital technology which I call the information revolution. Electronics can be identified almost completely

TABLE 3.5 *Cost of 3-minute call London–New York (in terms of hours worked by skilled manual worker)*

Date	*Hours worked*	*Technology*
1945	18	Short-wave radio
1956	12	Submarine cable
1967	6	Low-orbit satellite
1978	$1\frac{1}{4}$	Geo-stationary satellite

with the processing, carrying, and presentation of information. Whether it is radio and TV, radar and navigation, instruments and control, telecommunications, computing and accountancy, office systems and word processing, electronic typesetting and facsimile, information storage and retrieval, there is one common thread which runs through the lot: i.e. *information*. In electronic circuits, this information is best handled digitally rather than in analogue form. What is the difference?

Analogue form means the actual shape of the waveform, whether it is a sound wave or an electric voltage. When listening to a symphony, the analogue waveform is the variation of air pressure that hits the ear. It changes in infinitesimally small steps, and can assume any value whatever. It is easy to transmit, but also easy to distort; and once it is distorted, you cannot readily get the original waveform back – the output is no longer true to the original.

In binary digital processing, on the other hand, instead of an infinite variation in signal levels, we have only two signal levels to deal with. A signal is either 'on' or 'off'. If dealing with a string of figures, there is not an unlimited variety of figures, but only 0 or 1; and if you want decisions, you can only have two answers: Either Yes or No. As simple as that! The advantage of digital transmission is that any electronic device, as long as it can differentiate between two levels, will always reproduce the original number sequence, the binary sequence of zeros and ones that has been sent. Another point is that any analogue signal can be put into digital form. You may need very long numbers to describe a complicated analogue sequence; but basically every sequence, however complicated, can be put into a simple stream of binary figures, zeros and ones. It may, however, take a lot of electronic circuitry to do it, and until cheap semiconductor integrated circuits became available, digital processing had to remain a dream. Today it is no longer a dream, and we can use digital processing, not only in computing, but also for telephone calls, and for every kind of information transmission. Digital processing will permeate every aspect of communication and information, entirely because of the capabilities of the silicon integrated circuit to do it massively, quickly and cheaply.

Even video and sound recording will change from analogue to digital form in the near future, with the arrival of digitally recorded discs for the video long player and the compact audio player.

To show the impact of this combined silicon and information revolution, I show, in Table 3.6, a number of products which are commonplace today, but which were unknown in 1967.

TABLE 3.6 *Products in common use today – unknown in 1967*

Consumer products	*Image formation*
Pocket calculator	Thermal (infra-red) viewing
Electronic watch	Ultrasonic imaging
Video cassette recorder	Tomoscan
	Mini TV camera
Communications	*Computing and control*
Pushbutton telephone	The microprocessor
Electronic telephone exchange	
International direct dialling	
Personal paging	

THE MICROPROCESSOR AND ITS APPLICATIONS

The microprocessor is the key embodiment of microelectronics, because it is a cheap, compact, flexible tool which can remove the dividing lines between communication systems, information processing and control mechanisms; a tool for providing integrated electronic intelligence. The microprocessor is a true computer with the ability to store information, to take instructions, to calculate, to make decisions. But it can also be oriented towards specific functions in many of its applications. The microprocessor in the pocket calculator or the electronic watch, because its function is dedicated to a specific end, is not thought of as a computer any more. Because it is so cheap, you can use it not only for a complex role like running a super-tanker – but you can also do simple and trivial things with it in your own home.

Thus its applications will come everywhere: in the home, in the car, in the office, in the factory, in telecommunications, in transport systems, in harnessing energy resources and in food processing, but also in education, medicine, security, and all the service industries.

The range of microelectronics covers both existing applications by new methods, and radically new applications. One key to doing existing things by new methods is that many information devices still have complex electromechanical displays. These will be replaced; e.g. the hands on a watch are actuated by a precision mechanism that is quite expensive. The replacement is the all-electronic watch, without moving parts, and with a digital display. This is typical of the replacement of electromechanical information and control mechanisms by electronics. Another example is the mechanical distributor in the motor car, which will give way to all-electronic ignition control. Yet again, instrumentation with moving-coil meters, and electromechanical switching in telephone exchanges; all these are being ousted by microelectronics, which will be cheaper, more reliable, and more economical to run.

But where do replacements end, and radically new applications and products begin? Many products were unthinkable without microelectronics: the pocket calculator, the cash dispenser, the word processor, and the Teletext and Viewdata TV information systems will serve as examples. Let us look at some of these fields in more detail.

APPLICATIONS IN THE HOME

As an example, consider first electronics in the home. There will be many new uses for the TV set; these sets are now capable of giving not only a picture programme, but carrying information programmes of two distinct types, one coming over the air, known as Teletext; the other coming over the telephone, known as Viewdata. Teletext is cheaper because you do not have to pay for the information, but it is limited to a few hundred pages. Viewdata, where you dial up the wanted information over the telephone, will eventually provide thousands, hundreds of thousands, or even millions of pages of information. That is the starting point for the TV set as a home information terminal. Then it is only one step to a two-way terminal, where you can not only extract information but also put information into it, whether for a private user network or for public dissemination.

Then, there are the two kinds of video recording, the video cassette recorder and the video disc player. The video cassette recorder is essentially a time shift machine which takes recordings off the air when you do not want to see them, and plays them back when you do. The latest version will have 8 hours playing time, will pre-set 5 recording periods for up to 16 days ahead with freedom of programme choice. This flexibility is entirely the result of microprocessor control.

The linking of the telephone to the TV set paves the way to many new applications also linked to the telephone via a control equipment known as the home information terminal. One such application is telecommand, i.e. the remote control by telephone of domestic equipment such as cookers, washing machines, heating and ventilation. Another application is remote metering of electricity, gas and other supplies. Yet another is remote monitoring of security, environmental conditions or functioning of equipment.

The characteristic feature of all this is the fusion between the telephone, the TV set and the many domestic appliances, or more broadly put, the fusion between communication, computing, and control mechanisms. And if there is apprehension that too many important functions might be assigned to a single master-minding microprocessor, there is always the possibility of segmenting and decentralising this control with a number of individual microprocessors, since they will be cheap enough to allow this alternative.

In discussing the significance of all this I quote Professor John Westcott of Imperial College, London:

> Every new communication medium of this century, whether it was the telephone or the cinema, radio or TV, has elicited predictions of a new revolution that did not in fact come to pass; but the merging characteristics of 'phone, TV and computers will certainly represent something of a new order.

That is a restrained statement of what the strategic implications are all about.

APPLICATIONS IN TELECOMMUNICATIONS

Here, four main trends deserve mention.

(1) The *shift from analogue to digital* telephone signals, not only for the signalling circuits, but also for the speech circuits themselves. Digital transmission of many simultaneous conversations by time-division multiplex (TDM) demands very fast sold-state devices, as well as very high-capacity cable, but both these are now available; the cable will increasingly use the new technology of transmission by optical fibre.

Digital transmission with TDM will also allow the introduction of the new fully-electronic telephone exchanges, in place of the present semi-electronic exchanges which still use miniature reed relays. But even with existing techniques, international distance dialling already makes it possible to gain access without operator assistance to 360 million out of the total of 450 million subscribers in the world.

(2) The *merging between communication and information* processing has already been referred to under the heading of Viewdata. Its extensions to private business systems with multiple terminals are already under way. Further extensions, to include message storage, fund remittance, facsimile and electronic mail transmission, are not far away.

(3) The extensions of Viewdata just listed point the way to wider applications yet of the local telephone network. Tele-metering, tele-command, tele-monitoring are coming in the near future; and the suitability of the local network for transmission rates up to 64,000 bits per second allows even 'slow TV' to be sent into the home by telephone, without the need for wide-band cable installations.

(4) The modern electronic branch exchange has many new features made possible by microprocessor control – automatic recall of a busy extension; diversion of calls to other extensions; flexible routing through selected 'master' stations; conference networking. All these examples illustrate an important principle: the replacement of hard-wired connections by temporary software commands.

APPLICATIONS IN MANUFACTURING INDUSTRY

Here again, some important trends will be listed.

(1) The gradual replacement of present-day controls, whether electro-mechanical, pneumatic or hydraulic, by *electronic controls*. Since control mechanisms transmit information, not power, it is wasteful to use physical rather than electrical channels. The actuators, i.e. the mechanisms that translate the control commands into action, must of course retain physical elements to do their work, whether it is to operate a press tool, regulate fluid flow or shut off a furnace.

(2) Electronic control offers not only lower cost, but also added flexibility by allowing network configurations to be changed at will by different programming rather than by changed hardware connections. Here again, modern methods allow the alternative approaches of centralised or distributed control systems: and distributed systems lend themselves to piecemeal replacement of

older equipment. In the same way, fault diagnostics, so important in complex systems, can be centralised or decentralised at will.

(3) *Robots* have recently captured the public imagination, but I think they will have a limited place. Where precision manual operations need to be replicated again and again, such as the welding of a car body, an operation which is physically quite complicated but which is repeated from car to car, there the robot will come into its own and is doing so already, in activities like welding, de-burring of castings, component placement. Equally, robots have an important place in hostile environments of temperature, chemical vapour or radiation. Robots, as actuators, must retain mechanical or hydraulic components; but their often astonishing versatility should not be mistaken for intelligence, nor carried to uneconomic extremes. A robot that could collect dirty dishes from the dining-table, and place them correctly in the dish-washer, would be entirely possible but could not be cost-effective.

(4) *Computer-aided design and manufacture* (CAD and CAM)
The design of the complex electronic circuits mentioned earlier would be quite unthinkable without computer aids, because a large-scale integrated circuit may have tens of thousands, hundreds of thousands, or millions, of elements, many of which are replications of each other again and again. A draughtsman, or even a team of draughtsmen, in many cases could not do the layout, both because it would take too long, and because fatigue would lead to errors which might become undetectable. Computer-aided design of complex cell design tasks bridges these difficulties, and this goes not only for integrated circuits, but also for complete systems like telephone exchanges and computers themselves.

As a further benefit, CAM may help in making batch production economic again. You do not need large quantity production of a product if by electronic control it is possible to introduce a flexible degree of alteration of specification. For instance: the coding of the colour scheme of a car and its seats, whether it has a sunshine roof or not, and whether it has whitewall tyres or not. All of these are quite minor product variations, but they create a great nuisance in flow production because of the difficulty of controlling the information, the piece parts, the material flow. All this becomes possible again with computer-aided manufacture and re-opens the door for the product which is one of a family, which as a whole may have to be produced in large quantity, but where each variant may not need to be made in quantity. Computer aids in manufacture will thus be of importance not only to large firms, but to medium and small-sized firms as well.

APPLICATIONS IN THE OFFICE – THE WORD PROCESSOR

Next a brief glance at *electronics in the office – the word processor.* It is essentially rather simple, just a combination of separate technologies converging. The word processor combines several well-known units, such as a typewriter keyboard, a memory store, a visual display unit (VDU) an electric printer. The combined system is called a *word processor*, or *text editor*. Put the printer at the distant point and you have the beginning of an electronic mail system. Its uses are very widespread: text editing, text correction, document storage and retrieval, hard copies with small variance, such as similar documents with different addresses and openings, etc. When people express fears that word processors will lead to unemployment of office staff, it is in my mind totally misplaced, because they will be an aid rather than a substitute of office employment. But they do impose constraints, which will slow their general introduction. The biggest of these is that it is too expensive a machine for every individual secretary to have one. So it will impose an element of reorganisation in the office to ensure its most effective use.

Another constraint is that the storage of information in electronic files will only gain acceptance slowly. Most of us still prefer to have a filing cabinet with our Minutes, or with design specifications, safely tucked away in a file. However much we are told that electronic data storage is safe, secure and retrievable, it will be some time before we really are willing to scrap our filing cabinets. So the word processor will be a most important part of our offices, but in the middle rather than the very early 1980s.

APPLICATIONS IN THE MOTOR CAR

In the motor car the microprocessor will, in the next few years, have three important applications. It will be used in the control of ignition, in order to do away with the mechanical distributor. It will be used in control of the fuel mixture to achieve fuel economy, and will be used in control of the emission gases, to comply with legal standards that will undoubtedly be introduced. There will be additional uses which will come in as optional extras, like anti-lock braking, transmission and cruise control, suspension control, climatic control, novel methods of instrumentation and monitoring, and possible security and alarm controls.

There are some difficulties. Thus, the microprocessor cannot take over the control of the fuel mixture of air and petrol until there are sensors which measure what it is actually looking for; it needs a 'sniffer' which can turn its measurement into numbers that can be passed on to the microprocessor. These sensors, and also the actuators needed to turn microprocessor commands into action, are lagging behind in development. They will determine the rate at which electronics is really introduced into the motor car; then there is the fact that the motor car is a very hostile environment for delicate mechanisms because of its temperature extremes, its dirt, its vibration. So the full takeover of electronic control will take a few years, but when it does come the market for the microprocessor and its applications will be enormous. For the UK alone, car production will demand 5–10 million microprocessors per annum by about 1985.

APPLICATIONS IN NAVIGATION

Satellite navigation aids are already providing simple and accurate position indication anywhere in the world with the aid of the US 'Transit' system. Position measurement is intermittent at intervals of about 1½ hours, and interpolation is possible with any of the existing radio navigation aids such as Decca, Loran or Omega. The equipment gives position information with an accuracy of a few hundred metres, and its size has shrunk over the last few years to the dimensions of a shoe box.

A new system, the global positioning system (GPS), based on the 'Navstar' satellites, will become operational a few years from now. Out of a total of 24 satellites, at least 4 will always be in line-of-sight from anywhere. Coded signals from them will allow the solution of 4 simultaneous equations, giving x, y and z coordinates of position to within 10 metres, and time within one microsecond, anywhere and at any time. Indication will be automatic, and the equipment will weigh only a few pounds. It is an occasion for some wistful reflection that such developments will outdate navigational skills that have been built up over hundreds of years.

PROBLEMS FOR THE MICROPROCESSOR

In coming towards an end of this survey we must not forget that there are difficulties for the microprocessor too. The first, already mentioned, is the need for reliable sensors and actuators, i.e. the devices that pick up the environmental signals and the devices which translate the microprocessor commands back into physical action. These are still dependent on physical measurement and on mechanical power. The same is true in information processing. The input/output devices of information systems are still the weakest links – keyboards, tape-readers, displays, printers, are the weak and expensive parts to this day. In this field optoelectronics, a new technology alongside microelectronics, will play an important part in the improvement of input/output devices.

The biggest bottle-neck in the use of the microprocessor, however, is the need for software. As the microprocessor begins to play a part in every product field, software will join hardware as an equally important part of technology. A microprocessor is only as clever as the programme built into it. If we want it to do sophisticated things, it will need very sophisticated programmes. These take very skilled people, and will therefore represent a challenge to new skills on a very large scale.

SUMMARY OF STRATEGIC CONCLUSIONS

At this point some conclusions will be highlighted; most of these are summaries of remarks already made, but some of them taken a step further.

(1) *IC chip manufacture*
There have been great debates that the capability of a country to make its own microelectronic chips is basic to all its other industries. In my mind, chip manufacture is important but it can only be economic on a very large scale. It is beyond the powers of any firm based on a single European country to carry the necessary capability, with technologies becoming obsolete so quickly. Even a firm operating on a broad European scale, like Philips, has not got an adequate base in this field, unless it can link up with an American partner. Even in America, the displacement of traditional radio valve manufacturers, who set the pace at the beginning of the semiconductor industry, has been dramatic. Table 3.7 shows the 10 top tube makers in the world in 1950, compared with the 10 top IC manufacturers in 1979. Only 2 names figure in both the lists – RCA and Philips.

(2) *LSI technology* for integrated circuits will give great scale economies, but only if chips of standard form are used in adequate

TABLE 3.7 *Top-ranking manufacturers*

Radio valves (1950)	*Integrated circuits (ICs) (1979)*
1. RCA	1. Texas Instruments
2. Sylvania	2. Philips/Signetics
3. Philips	3. National Semiconductor
4. General Electric	4. Motorola
5. Raytheon	5. Intel
6. Westinghouse	6. Fairchild
7. Amperex	7. Nippon Electric
8. National Video	8. Hitachi
9. Rauland	9. RCA
10. Eimac	10. Toshiba

numbers. This does not exclude custom-built chips, as long as the numbers are large. The same remark applies even more strongly to VLSI technology. Layout of these circuits and design of their test routine will be very expensive, and may therefore become self-defeating except for very large-scale use of a circuit. Storage is the obvious standard application.

(3) *Microelectronics and microprocessor applications* are vital to every industry, not just the electronics industry, but in every phase of manufacture, processing, transport, communication, energy supply services, home entertainment, and domestic equipment. Failure to incorporate the microprocessor in products or in processes will be fatal to competitiveness.

(4) *The new information technology* enables both the realisation of very *complex systems*, e.g. the world phone network with 450 million subscribers, but also very *decentralised organisation*, where processing and control in big computers can be replaced by distributed processing, decentralised computing, decentralised controls. This in turn means that the new technology makes possible variations of product series, or product runs, variations of customer choice, and a re-emergence of batch technology. The dominant result of the use of intelligence in industrial processing should be to provide additional flexibility and personal involvement, i.e. *not* to exclude the human being from retaining control. Automation should therefore have a positive impact on work quality and on work interest, and on work satisfaction.

(5) It follows that it is not only in the large factory where all this becomes important, but in the home, in the motor car, in the small

factory, in the small office. These are the most fruitful fields for the application of the new information technology. The small office will have the choice of its own stand-alone computer for a few hundred pounds, or it can hire a segment of a computer utility with access over the telephone network.

(6) *Software*, i.e. the organisation and programming of the intelligence of systems, is becoming the most important, the most expensive, and the most difficult aspect of the new information technology, and will demand an ever-increasing proportion of engineering resources.

(7) After software, the biggest constraint in the spread of microprocessor applications are sensors and actuators in control systems, input and output devices in information systems. A lot of work needs to be done on all these, to get their quality, reliability, low power consumption, and low cost to the level that microelectronic components are achieving.

(8) On more philosophical ground, we must not allow information processing to become a substitute for information generation. The *creation* of knowledge, of intelligence and information, and its spread, is something quite different from electronic information *handling*. No machine can do that, and creative initiative cannot come from a machine.

(9) *Innovation* is essential, but it also implies obsolescence – the willingness to discard old techniques and products. It implies risk:

    the willingness to take risk;
    the ability to make decisions;
    the ability to pull out of failures;
    the ability to pool knowledge with others.

These are sobering thoughts. Robert Noyce, who may be regarded as the inventor of integrated circuits, in 1977 put on record, 'the environment for entrepreneurial innovation in the US is not matched in other industrialised countries'. This is a moderate statement which is difficult to contradict.

What happens in Europe? We distrust the governments of our countries. We resent government interference, and we certainly resent the taxation levels under which we operate. Yet we run to our governments for protective tariffs, for subvention of our risk investments, and even for technical policy initiative.

If European industry is to survive the challenge of the new technology, it must not abrogate to government agencies the initiative of creative innovation, design ingenuity, risk investment or marketing enterprise.

(10) *Employment.* The new information technology may become both a threat and a stimulus to employment. Many traditional activities will shrink in importance, especially if concerned with the manufacture of electromechanical control gear, but demands for new skills and even new industries will persist. This poses problems of retraining and of mobility which are by no means trivial.

The great value of the new information systems is in carrying out functions that were either not performed at all, or not on the same scale formerly. Every new system can bring its own requirement for new people. Exactly as the computer itself created more employment rather than less, so the new generation of office aids, the word processor and the electronic information systems, will enhance the need for office work. The nature of the work will certainly change, but every pointer is that it will be more varied, more interesting, and will call for more intelligent involvement.

The new information technology should create growth, wealth, and new jobs. But there will need to be much flexibility redeploying existing skills, and retraining people, so that they get used to the idea of pursuing more than one type of skill during a working lifetime of 40 years. Only the combined ingenuity and resourcefulness of our managements, of our labour forces, and of our public and government agencies can make these changes successfully.

4 The Effect on Employment

RAY CURNOW

Historically, technical changes and the resulting technological developments have always resulted in a remapping of the pattern of economic activity. Nationally and internationally, the location of industry has altered to fit new practical requirements; and structurally, the interrelationship of supporting and dependent activities has varied with the nature of the central development.

The wide range of impact of even a limited-scale technical advance can be seen in an event like the mechanisation of the textile industry at the beginning of the Industrial Revolution. Geographically, the woollen industry moved from the South-West and East Anglia to the growing new towns of Yorkshire. The cotton industry came from farther afield: while work became available in the Lancashire mills, there was an enormous negative impact upon the indigenous spinning and weaving industries in cotton-growing areas like India.

On a micro level, there was a shift from work in the home towards work in factories, while on a macro level world trading patterns altered, and the level of trade grew, to accommodate the new pattern of activities. The repercussions spread outwards in waves, first affecting related industries like dyeing and tailoring, and eventually becoming inextricable from the broader movements of the developing revolution. New technical changes grew in parallel with the original upheaval: the development of synthetic dyes in Germany, for instance, which formed the basis of the European chemical industry, and sounded the death knell for the third world production and export of natural indigo.

Whether the effect of a change such as this upon employment is, on balance, good or bad, depends very much upon where the boundaries are drawn. Within an economic subsystem, the picture will emerge in one way: make the radius of the circle larger or smaller, and the scene may change entirely. The growth, levelling off and gradual (or not so gradual) decline of a dominant production process has been well documented, and it is generally recognised today that the single-product industry is extremely vulnerable to obsolescence through technical change.

Mankind has only a limited selection of needs and desires, and to a great extent these can be satisfied by a variety of products, produced in a variety of ways. The march of modern technology has made many goods and services available to the population of the world. Not just the elite, but a large proportion of the global population enjoy luxuries unknown to their forefathers. At the same time, it has led to the decline of other goods, and other methods of manufacture. This is particularly true in terms of employment, where the shift from agricultural employment to factory work, and from there to the service sector, has been well recognised.

Microelectronics is no exception to these rules, and the impact of microelectronics stands at the centre of an interlocking web of changes and developments. Ripples radiate outwards from the stone thrown in the pool: but the pool contains many other movements, and the pattern of the ripples is easily lost among the patterns caused by other flows and obstructions. Geographically, the impact of the new technology is inextricably linked with the wider trends of world economic activity and international trade. Technically, the contribution of microelectronics is intertwined with that of other technical developments such as lasers and fibre optics.

In considering the effect of microelectronics upon employment, I shall divide my analysis into two parts. The first will review a 'Robinson Crusoe' economy: that is, the impact in an area in which the effects of trade and competition are either ignored, or encompassed through scale. In my second part, I shall consider the realistic effects within a competitive economy.

What, then, is the isolated impact of microelectronics? It is already clear that microelectronics is not a limited technology. Its uses and applications, directly or indirectly, affect every area of the economy. From primary production to sophisticated software programming, in factories and in offices, the microelectronic revolution is taking place, and a corresponding effect upon employment will become apparent in all these areas.

Perhaps most simply, microelectronics can be used in direct substitution for previous electronic circuitry carrying out the same function. The silicon chip merely replaces a larger integrated circuit: in computers, hi-fi systems, production control machinery and other electronically controlled equipment and appliances. This substitution may also pave the way for the adoption of associated technologies made available through microelectronics: fibre optics, for example, which are having an enormous impact upon telecommunications, or laser applications.

As well as replacing existing electronic circuits, microelectronics finds new applications in areas where electronic circuitry was not previously applied, either through technical constraints, or due to price restrictions. New goods such as television games are appearing; new control machinery such as point-of-sale microelectronic terminal/cash registers, and new production processes and applications – involving advanced robotics, for instance. Incorporation of microelectronic capabilities in revised versions of existing goods – 'intelligent' door locks, memory typewriters, programmable ovens – reveals another aspect of this process. Similarly, new applications are opening up for the associated technologies: laser-controlled holography, with potential usages including high-density storage and three-dimensional advertising material, is one example of the expanding fields.

These new goods and service opportunities may provide for some additional employment openings, and this is certainly consistent with the pattern of previous technological revolutions. Just as Lancashire and Yorkshire became boom areas in the early nineteenth century, so we are today seeing a boom in the more sophisticated cheap-labour countries like Taiwan and South Korea, in Japan, and in 'Silicon Valley', California.

An analysis of the effect upon employment must take account of the birth of new jobs, the life-cycles for new products and processes. At the same time, it must consider the death of old products and processes, brought about or hastened by the changes. The net effect on jobs will, needless to say, be the difference between these two opposite but related developments.

To assess this effect, it is necessary to undertake a remapping of the job pattern: to review the evolving processes of economic activity, and to structure an analysis of the future job market according to this total pattern of gains and losses. However, it is perhaps more feasible to begin with the existing production and employment infrastructure, and to assess the changes that can be anticipated here. Newly perceived opportunities can then be mapped onto this revised and projected framework.

The production processes which provide many existing jobs are being altered with the advent of microelectronically controlled production machinery. This process might broadly be summed up as the development of automation. In some areas the actual production jobs are being automated. The result here is that assembly-line workers, paint sprayers and many other direct production operatives are being replaced by robots. In other areas it is the control of the production process at a factory level that is being automated, with the aid of sensors, gauges, 'vidicon' cameras and other equipment tied into a computerised control system. In this case, it is production control jobs that are at risk.

Both direct production jobs and supervisory jobs are being lost, in this way, to microelectronics. Many factories today are manned by a skeleton staff, while virtually all routine production activity is carried out automatically.

Products in which silicon chips replaced existing electronic circuitry are becoming simpler to make, and this effect is even more apparent in products where the chip replaces existing mechanical works. Typical examples are the wrist watch and the calculator: digital watches and electronic calculators contain hundreds, even thousands, less parts than mechanical watches and calculators. The work involved in assembling these new products is obviously a great deal less than that formerly required, and even where the assembly is carried out by hand (the increased simplicity of the operation again encourages automated assembly) there will be, or indeed already is, a loss of jobs in these areas.

The energy required for this abbreviated production process is equally obviously reduced, and so, in many cases, is the quantity of raw material employed. The production base of the economy might be thought of as a pyramid, divided horizontally into layers representing the proportion of work (both production and employment) involved at each stage of the manufacturing process. At the base of the pyramid would lie the task of gathering and preparing component blanks; and on the next levels, the fabrication of components, the assembly of subsystems, the assembly of the finished product, installation, testing and maintenance, and finally, at the tip of the structure, the design and marketing of new products.

In effect, under the impact of microelectronics this entire pyramid shrinks in size: less raw materials are used to make less, and often smaller, components; subassembly and final assembly are simplified; and even installation, testing and maintenance become less labour-intensive operations. The peak of the pyramid is not immune from this process, either. Design of new products is proving to be an early area for automation, with computer-aided design techniques augmenting traditional design engineering skills. Again, in the medium to long term, this will lead to a reduction of available jobs.

Finally, administrative and support functions are not exempt. Data processing has been a reality for a long time, but microelectronics is making it much cheaper, and its uses correspondingly wider. Word processing is now economic for a wide variety of applications, and these two automated functions are cutting a swathe through the ranks of office staff. With the advent of integrated information technology systems, combining processing (of both text and data), storage of all kinds of information, and, through the augmented capabilities of telecommunications, rapid and effortless transmission and communication of this information, the process should continue to accelerate. This will, in turn, have a more than peripheral effect upon sales and managerial staff, as the 'pyramid' of clerical support staff shrinks beneath them.

Sales and marketing operations are also subject to the microelectronic revolution. At a retail level, the introduction of point-of-sale terminals is speeding checkout operations, and providing an automated stock control system. Discount warehouses are growing rapidly upon the rationale of minimum staffing, and customer self-help supplemented by computerised warehousing operations. 'Broking' and agency operations, selling financial and other services such as travel and insurance, will be vulnerable to the spread of public database services incorporating direct ordering and payment facilities, while the 'intelligent credit card' and electronic funds transfer, together with autobank facilities, should combine to have a revolutionary effect upon the broad financial sector.

The control aspects of microelectronics could also have an effect upon the production of utilities. Marginal pricing, differentially computer-controlled according to the varying levels of demand, could help to flatten the peaks in the demand curves for gas, electricity, and even public transport, and so increase the utilisation of existing resources, and reduce the demand for new.

Regarding new products and processes – though new and improved products made with the aid of, or even incorporating, microelectronics appear in our shops almost daily, it is difficult to identify many of these which fulfil totally new functions. New production machinery replaces old; new components old; new products old. Almost the only completely new microelectronic product is the television game, almost certainly a short-term phenomenon, as these capabilities will soon be incorporated into the standard television set. Even then, this product is taking custom from other toy and game manufacturers.

Cheaper microelectronic products can achieve greater market penetration than their mechanical predecessors – far more people own an electronic pocket calculator today, for instance, than used to own a mechanical desk calculator. But even this vastly improved level of sales is not echoed by similar improvements in production and employment figures. Turnover of the pocket calculator industry today is lower than that of the old desk mechanical calculator industry: the increased volume of individual sales is not an increased cost volume of sales.

The pocket calculator industry also employs less people. There is a particular economic incentive to automate the production of new products, where no investment in more labour-intensive machinery has already been made, and it is difficult to imagine a totally new product which would not be produced by a highly automated method. Even if the new product did not replace existing products to a greater or lesser extent, it would not be likely to lead to a substantial creation of new jobs.

Growth of demand, in any case, is not purely a function of theoretical price elasticity. It depends upon purchasing power, and the distribution of disposable income. The increase in levels of unemployment, reducing the purchasing power of a large section of the population, may well lead to an adverse feedback effect upon demand for new, even relatively cheap, products, as a technologically induced recession begins to bite.

Within the wider horizons of a truly competitive economy, yet other factors come into play. With the severing of the links between labour and production, the real message behind the microelectronic revolution, the door is open to a total relocation of economic activity. It is now perfectly conceivable that worldwide production of all electronic goods, for instance, could be concentrated within one country, with Japan as the obvious prime candidate.

The increasing capabilities of information technology have a major part to play in this revolution in economic factors affecting location. With increased accessibility of information, the worldwide market is becoming more and more of a reality, and its perfectibility is increasing steadily. Inevitably, international competition will become more intense, and less room will be left for the less efficient manufacturer or provider of services.

This ease of communication may also affect the pattern of location within an individual firm. Multinational corporations may opt to place their production facilities in third world areas where land, labour and overheads are cheap, and to retain virtually all of their white-collar support staff (managerial, data processing, research and development, and so on) in developed countries. The resulting 'information gap' could make it even more difficult than it is today for third-world countries to attain an independently higher level of development.

While economies of scale have provided an incentive for the larger company in functional areas where large-scale production is practicable, there has, up till now, been a large 'fringe' area, between the stereotyped production of the multinational and the individual product of the craftsman, where the small to medium-sized firm proved to be the most efficient. Microelectronics is lending a new flexibility to large-scale production equipment, making the task of changing size or other specifications much simpler – for example, in clothing manufacture, or heavy engineering. This increases the advantage of the large, highly capitalised firm, and may provide a further squeeze for the smaller firm, in which the production processes are often more labour-intensive.

The enormous advantages that microelectronics provide inevitably mean that rejection of the new technology (unless a siege economy is to be the result) cannot be the answer. The competitive advantage enjoyed by those firms and countries which adapt and adopt the technology will be even greater, and the net outflow of jobs to more efficient producers may prove even larger than the likely job loss through automation.

In this manner, job loss within the current sectors of major employment would seem to be inevitable. Within a worldwide picture, less labour will be required to produce a given supply of almost any good, and all but the most personalised services. The geographical distribution of this reduced labour requirement is the main question open to determination in the period in which adaptation is taking place.

Elasticity of supply might in theory open up new employment opportunities: but in practice it is difficult to identify many, if any, goods for which the expansion of market demand outpaces the expansion of opportunities for automation of production. The obvious conclusion is that the present-day decline in employment in the manufacturing sector, and in large parts of the service sector, is as inevitable as the decline in agricultural employment that resulted from the automation of agriculture and the coming of the first industrial revolution.

The impact will clearly have a differential effect in different occupational sectors: but there are very few occupations where it will be negligible. In the short to medium term, the major effect would appear to lie in the field of semi-skilled and skilled labour. A high proportion of factory jobs within this broad classification are ripe for automation, and so are a sizeable number of clerical and secretarial jobs of a comparable level of complexity.

Automation lends itself to the fragmentation of processes, and the pockets which escape the first waves of automation are likely to provide menial jobs: factory or street cleaning, for example, and other repetitive jobs for which robots cannot easily be dedicated. Similarly, there will be a residual proportion of 'back-up' jobs, waiting for snags and breakdowns in the predominantly automated processes to occur. These tasks are likely to prove unrewarding, with a boredom quotient at least as high as that of many unpopular factory assembly jobs today.

At the upper end of the job market, the impact of microelectronics upon professional jobs will be of a different order. The advent of relatively cheap microprocessing and data storage facilities will make these more readily available (even on a one-to-one basis) to professionals such as doctors, architects and engineers. Similarly, the development of more 'user friendly' computer peripherals should encourage specialists in other fields to make use of these facilities in their daily work. The consistency of the well-programmed computer can lead to its outperforming humans, not merely in number-crunching activities, but in broadly defined 'information processing' areas such as diagnosis of illness. In this way, many of the more routine aspects of the professional's job, those involved primarily with the application and processing of previously memorised information, will be subject to the impact of the microprocessor, and there may develop more of an emphasis upon human conceptual and imaginative skills. Some professionals – design engineers and architects, for instance – may experience a shrinking in their overall work field. Others will remain in high demand, but be subject to qualitative changes in job content, in many cases for the better.

A similar pattern may emerge within the computer field itself. While the conceptual skills of the systems analyst may be in great demand, it is conceivable that there will be widespread automation of the routine programming and computer operation tasks which make up the bulk of computer-related employment.

The overall effect of these processes may prove to be a movement towards a 'two-nation' situation. There could develop an unbridgeable gap between the machine minders and the machine makers, between very low-level and very high-level occupations, with few jobs remaining in the 'middle' range of skilled manual and lower-grade white-collar employment. The implications of this could be disastrous. Not only might education and training of this divided work-force prove a nightmare: it is also difficult to imagine how such a two-tier society could maintain a democratic social structure.

Many 'service' sector jobs fall into this middle range of employment, and it is perhaps to these that we should be looking for the development of new employment opportunities. It is an over-simplification, however, to suggest that the shift in the period of the first industrial revolution from agricultural employment to manufacturing employment will be paralleled by a new shift from production employment to service employment.

A reassessment of the catch-all service sector leads to the conclusion that this generic term encompasses two very different areas of employment. Firstly, there is what might be defined as the 'practical' service activities: that is, both personal service (domestic services, hairdressing, restaurant and bar work, taxi driving and so on) and public service (teaching, social work, bus and train driving, etc.). And secondly, there are activities which are primarily concerned with the handling of information: administrative functions within local government, banking and insurance, brokerage trades (financial and insurance, but also travel, estate agents, and so on).

A redefinition of this second subsector as the 'information sector' can also lead to a reassessment of the manufacturing sector in the same light. Support operations like secretarial work, accounting, and data processing have traditionally been defined and classified according to the product of the firm for which they are carried out. These too are, however, primarily 'information sector' occupations, and their inclusion within the figures for manufacturing sector employment has the effect of artificially inflating the numbers of those employed in manufacturing.

Recent analyses (Barron and Curnow, 1979) indicate that employment figures could be allocated within this four-sector classification as follows: information sector, approximately 40 per cent; industrial workers, approximately 20 per cent; service industries, approximately 36 per cent, and agricultural workers, approximately 4 per cent. Further analyses (Porat, 1976) show that, as one might expect, it is the information sector that has shown the major growth over the last 100 years. Service sector growth has been only modest, with the boom in public service operations largely compensating for the decline in the number of domestic servants.

It is not by accident that microelectronics and its family of associated technologies are becoming generically known as 'information technology': for it is upon information handling, storage and transmission that their impact is centred. It is to be expected, then, that major changes in this essential information sector will result from these technological developments.

In general, the information sector acts as a microcosm of the economy, exhibiting the same pattern of growth, change and decay in individual activities that has been identified in the economy as a whole. In this case, newly developing activities include word-processing (the design, implementation and operation of both small-scale applications depending upon a stand-alone processor, and large-scale projects using a pool of more advanced equipment), videotex (Viewdata, Teletext and associated developments) and new printing processes such as ink-jet printing.

The employment generated by these new processes and operations is likely, however, to be outweighed by that lost through the decline in traditional information sector occupations. Routine and semi-routine clerical operations, typewriting, manual filing, manual switchboard operation and the various kinds of traditional printing processes exemplify areas in which employment is either threatened, or already in relatively rapid decline.

The capacity of bureaucracy to proliferate, and to protect itself when under attack, should never be underestimated. It is already becoming clear, though, that the development of the information sector as a whole has reached its plateau stage: and a decline, the speed of which will depend largely upon broadly political factors, is to be anticipated.

It is thus apparent that it is futile to look to this sector of the economy for an expansion of employment opportunities. Any potential for new employment must be sought in the service sector proper.

Ironically, current economic policies are resulting in a squeeze upon public sector service employment: a policy that would appear short-sighted, since even an exceptional performance in manufacturing will not obviate the need for many more jobs to be found in the near future in non-productive areas.

While there are, of course, opportunities for automation in some service occupations (teaching, for instance, could well be transformed with the aid of computerised aids to learning; the newer underground transport lines require relatively few staff; and there are many more examples available) it is still true that much manpower might desirably be redeployed to the 'social service' areas: the health service, care of the old and disabled, and the more general community care services. Certainly this is a more desirable future scenario than an uncontrollable growth in unemployment.

This kind of political choice is one of the imponderables that make it impossible to predict how much unemployment will result from the changes in technology. We can estimate the potential effect upon current jobs; but not the way in which unions and government will combine, or fight, to produce the actual effect. And nor can the effect of long-term creation of jobs be quantified in advance.

An estimate of potential job losses, based upon current knowledge, leads to the scenario of an overall job loss of between 7 and 20 per cent over the next fifteen years (Barron and Curnow, 1979). This would be predominantly in information handling and production areas, with the impact upon agricultural workers being least.

Just as the impact will differentially hit types of work (and, indeed, different specific jobs), so it will differentially hit sectors of the working population. Experience to date with unemployment has indicated that those who suffer most are the young, and immigrant populations. The disturbing figure that for every 1 per cent increase in the unemployment rate, there has been an increase about eight times larger in the unemployment rate amongst black youngsters (observable from *Social Trends Statistics* (HMSO)) suggests that serious social disruption might result well before the level of impact envisaged above is reached.

Young people are hit, not only because of their inherent unattractiveness to employers, but by union policies on redundancy. As more and more unions and employers sign 'no redundancy' agreements in areas subject to declining labour requirements, so these areas are effectively closed to new entrants. Already a pool of young people is growing up who have never had a proper job, and as time passes their aptitude for work of any kind decreases almost as fast as their 'saleability'.

At the same time, it is difficult to find an incentive to encourage these youngsters to make more effort at school. For all but the brightest, the long-term prospects for employment are little better if they obtain educational qualifications than if they do not. Though employers can afford to pick and choose (and, in some cases, demand 'O' level qualifications for menial dead-end jobs) they are likely to find that ambitious young people, confronted with crumbling or non-existent career patterns, rapidly become disenchanted with the working environment.

Particular problems may also confront women. Two of the largest areas of female employment have been light assembly work in factories (often on a 'twilight shift' basis) and office-based sacretarial and clerical work. The former area is disappearing with the supercession of electromechanical products by simpler microelectronic versions, and the development of robotic assemblers. The latter is being decimated by the introduction of word processors and the microelectronic 'paperless office'. For many women, their work is not merely for 'pin-money'. A growing proportion of women are the sole support of one-parent families, and the contribution of many more helps keep the families of low-paid men above the poverty line. Women in this position cannot be expected to surrender their jobs willingly, unless that surrender is accompanied by a massive shift in social security provisions.

The picture painted here is black, but it does not seem unrealistic. Though our economy has, historically, coped with upheavals on a similar scale, it has not done so without an enormous amount of individual misery, and nor has it done so in a comparably short time-span. The speed of penetration of microelectronic technologies is already outpacing the early forecasts, and the higher level of utilisation even today in the most advanced countries (Japan and the USA) suggests that we have not yet seen the end of this process.

Whereas the first industrial revolution was the source of the division between capital and labour, the present revolution is threatening to sever all the remaining links between these two elements. Up to this point in history, a proportion of labour has always been an essential ingredient of every manufacturing process, even the most capital-intensive. This is rapidly ceasing to be the case. We are reaching a stage at which we can envisage creating all the goods we require (supply of raw materials permitting) with the aid of only a skeleton work-force. The rest of our historical work-force are thus disenfranchised of their stake in the wealth-creating production processes, while the role of capital becomes all-important.

The cost of providing a job in modern manufacturing industry has become so great that, even were it desirable to do so, the amount of capital available in this country could not be stretched to provide one for every member of our historically constituted manufacturing work-force. Few would argue that, in a world where the limits to growth are already all too apparent, this should be our course. The only alternatives, though, are a radical rethinking of the nature and role of work in our society, and a probable collapse into chaos.

Our traditional checks and remedies no longer appear sufficient to prevent the emergence of such chaos. Most trade unions are adopting a cautiously positive approach to the new technology, but their first loyalty is to their members and their members' jobs, and in the circumstances they have little alternative but to fight a rearguard action: accepting technological changes only when guarantees of no redundancy are given. These can be achieved, in most cases, only by an exceptional increase in production, or by the acceptance of deliberate overmanning – as the management of *The Times* have found to their cost.

Overmanning, though perhaps a necessary evil within the bounds of our national economy, can prove disastrous for international competitiveness. Even in these circumstances the remedy is only partial.

The inevitable turnover in employees, and the steady 'natural wastage' of a fixed work-force, weaken the checks and swell the ranks of the unemployed.

A more positive approach to the problem might be oriented around a consideration of the nature of work. What are the values of work, and what are its demerits? Is there work that, nationally or internationally, is genuinely best done by machines and microprocessors? And is there work that should, for moral or idealistic reasons, be reserved for man? This may sound ingenuous, but a radical reassessment of our long-term goals provides the only possible sound basis for the kind of radical restructuring of our society, and the place of work within it, that is coming to appear not merely possible, but unavoidable, if our society is to survive the microelectronic revolution.

What, then, is the function of work? It is, in our current society, a means of obtaining a preferred income. This is not, however, its only value. Work, for most of us, provides our cards of identity: we introduce ourselves by indicating what we do for a living. It provides a structure on which we base our social perceptions, and our pattern of life, and for many it is a major source of socialisation – we make most of our friends at work. Finally, and not to be underestimated, it is a means of occupying our time.

It should not go unappreciated, either, that work has negative aspects. Many jobs are routine and boring: workers endure them, rather than look forward to them, and in some cases actively detest them. Others are dangerous (coal mining, deep-sea diving, and working with asbestos, for instance) and thoroughly unpleasant (paint spraying, and many other factory jobs carried out in hot and dusty or fume-laded conditions). These are by no means an essential part of a civilised society, and there would seem to be a real merit in developing mechanised ways of carrying out these tasks. Many people whose jobs are neither dangerous, boring or overtly unpleasant nevertheless dislike them, or can think of many other things they would rather do with their time. And again, many people with young families or dependent old, sick or disabled relatives are sufficiently burdened without the additional problems posed by paid employment.

It is one of the paradoxes of our society that we should both look forward to leisure, and fear unemployment, while both these terms refer to fundamentally the same phenomenon – the absence of work. If there is to be less work available in our society, it would seem to be clear that periods of non-work should be planned so as to resemble leisure, rather than unemployment.

One fundamental requirement here might be to distribute this valuable free time more evenly. The necessary corollary of this would be a more equitable distribution of available work, both by discouraging those who have too much work to do, and by encouraging the employment of those without paid work. This new free time/work division might be accomplished, for instance, by reducing the working week (either shorter working days, or less days' work per week), increasing holiday allowances, encouraging 'sabbatical' periods for long-term educational or travel activities, or lowering the retirement age.

It should be remembered that neither our definition of paid work, nor our concept of the normal working week, is an immutable fixed point of reference. Sixty per cent of the population, for instance, do not participate in the activity of paid employment: it is the arrogance of those who do which leads them to speak as if for the majority. Our concept of a fixed working week derives largely from the time of the Industrial Revolution, and the division of our waking lives into 'work' and 'leisure' periods from the same time. Many cultures do not have any such concept, for the distinction relates primarily to factory and office work: it is alien, too, to a high proportion of the self-employed. Similarly, 'unemployment' is a culturally based concept: in both primitive and socialist societies, it is one that has no meaning.

One of the great opportunities opened up through microelectronics is a dissociation of the activity of working from its traditional locations. Just as (as explained earlier), companies have the opportunity to scatter their global activities, so might individual employees alter their place of work. Obviously a manual worker has no such choice, but most of those in the information sector may soon be able to function away from their present-day offices. Telecommunications and the development of remote computer terminals means that any location with a terminal facility can obtain full access to database and communication capabilities.

This phenomenon might be seen as a logical development from today's correspondence courses and home-based computer programming bureaux. Professionals in many fields, from accountants to teachers, word and data processors, technicians and managers would all be able to spend at least a proportion of their time working from home. Though available for urgent consultation when required, they could also be free to carry out home maintenance, child-minding and leisure activities in intermediate periods. Though such a process might bring some new participants into paid employment, these could well be outweighed by those currently working full-time away from home, but who would prefer a greater flexibility in their working lives.

In other areas, the barrier between formal work and other types of activity is already crumbling. The disparity between the supply of labour available, and the demand for labour at an acceptable wage level (allowing for our disproportionately heavy taxes upon labour), has led to a healthy 'black economy' in which those already adequately supported by first jobs or by social security payments make their labour available at a lower, untaxed price. Estimates of the size of the black economy vary widely, but it is clear to the most casual observer that it is far from insignificant in a wide variety of service areas and labour-intensive manufacturing/repair/maintenance activities.

There is a particular incentive for the formally unemployed to supplement their base income in this fashion, and an official 'blind eye', accidental or deliberate, does much to alleviate the stress in areas of high unemployment. Doubtless a few electronics buffs are already using silicon chips in their garden shed workshops, but for the most part the impact of microelectronics upon the black economy will be more tangential, working through the wider forces of unemployment and wage levels.

Where personal services and 'good works' are involved, it is the dividing line between paid and voluntary employment that becomes blurred. The black economy fades into the barter economy, into baby-sitting tokens and illegal child-minders, neighbourhood co-operatives and car pools, while the paid social and public services blur into voluntary services, charities and community associations. If these activities are not part of the economy in the sense that they increase production, they are a part of working life, and an expansion of this area not only provides employment for those redundant in the manufacturing sector: it also helps to improve the overall quality of our lives.

Summing up at this stage, it is clear that the employment issues for management are three-fold. The pattern of employment controlled by management will change, certainly in quality if not in quantity; the management role itself will evolve likewise, and probably more rapidly, and finally, the external environment of employment, both formal and informal, is subject to change. The extent and nature of these changes is not yet clear, and certainly will vary enormously between different functional areas. It is doubtful whether the macroeconomic trend indicators of productivity, capital–output and capital–labour ratios will remain as useful trend-indicator tools as they have been in the past. Given that the new technologies embody capital in a way which is, in general, both capital-saving and labour-saving, that the adoption of such new technology is probably at the behest of the imperative of competitive need, and that it is beset with socio-political consequences, the relationship between macroeconomic and microeconomic behaviour will become even more tenuous.

To illustrate these points, it may be useful to consider again the structure of the UK economy. The normal and traditional classification into SIC categories, i.e. by nature of output, may not be as helpful as one crudely organised by economic, or even social, control. Such a classification might usefully be along the following lines:

(1) Transnational companies, operating in competitive world markets, both manufacturing and commercial. A further illuminating subdivision might be according to the degree to which physical investment ties such companies to the UK, or the relative weight of the UK market to their other interests.
(2) UK-owned companies, but largely competing in world markets.
(3) The nationalised industries: rail, coal, steel, transport, and utilities.
(4) UK companies solely concerned with satisfying UK markets – a further subdivision might consist of those who serve more local markets, or owing to the nature of their productive factors are relatively immune from import competition.
(5) Central and local government activities.
(6) The large centrally directed services – for example, the National Health Service, Department of Health and Social Security, the Inland Revenue, and the educational sector.
(7) The 'black' economy, and other fringe activities.
(8) Population-based infrastructural activities, e.g. distribution.

Although some companies would not fit easily into this classification (the major UK banks, for example) it does bring out the different sorts of pressure, and response to pressure, open to different types of economic organisation more clearly than more traditional classifications. As is well known, some organisations can absorb low productivity or high wage increases by passing on costs to the next consumer, some organisations are more footloose than others, some are subject to governmental mandate, and the degree of unionisation varies considerably. It would appear, though, that a discussion and analysis of occupational shift within each of these categories is more likely to give an indication of employment trends than are the methods employed to date.

Yet overhanging even such a detailed analysis are larger political and economic issues. As a trading nation, the UK is heavily dependent, though no longer a dominant agent, on world trade. In this context, the future of the US economy is still of paramount importance; and yet the major question marks about that future still concern the impact of oil prices and the US agricultural surpluses – subject on the one hand to gross political uncertainty, and on the other to the impact of weather on the demand for cereals.

The dramatic switch in UK government policy following the 1979 election has not yet had its consequences. One view might be that it will hasten the day when the UK becomes either the Argentina or the Poland of Western Europe: a non-growth society, with a dual economy of multinational investment and a stagnant socioeconomic framework in other sectors, or a highly bureaucratised and relatively closed society. Such a choice would doubtless be hotly denied both by zero-based government proponents on the right of the Conservative party, and by siege-economists on the left of the Labour party. The task of re-energising UK industry and commerce which faces both factions has, though, been made more extreme and urgent with the arrival of the powerful new information technologies.

Within the UK itself it is particularly unfortunate that the second post-war population bulge is now beginning to enter the labour force. At a time when employment is under pressure, many apparent remedies have the effect of restricting new entry routes to the formal labour market. The trend to decreasing long-term youth employment has been clearly visible for several years and presents a structural division within society which may have dire social consequences. In the same way, the geographical distribution of the older declining industries and the newer industries is such that economic well-being is now to a large extent mappable by region.

The extent to which political leaning is now associated with location is unparalleled since the 1930s, and this may also be a major social issue, particularly in a society where the private, as opposed to the public, costs and benefits associated with the location of economic activity are distributed by different, and often opposed, mechanisms.

For many years economists have argued over the existence of Kondratiev cycles (Kondratiev, 1925). Broadly speaking, Kondratiev, a Russian economist working in the early Soviet Union at a time when the acceptance of such ideas was a political issue and therefore variable, argued that industrial history had already seen long waves of recession following major new technologies. In crude outline, both the coming of the railways and the development of electrical power were characterised by a spurt of exploitation, followed by labour-saving investment and an associated slump. Some, myself included, would argue that the recent world recession, which certainly began prior to the oil crisis of 1973, is but the beginning of a Kondratiev surge following the unparalleled research and development activity of the 1960s, and that microelectronics is just the tip of an iceberg of technical change yet to be absorbed. If the Kondratiev mechanism is accepted as valid and still applicable, then the recession would, in today's more interrelated world, be both deeper and longer.

Although this theory is highly contentious, since the data supporting so-called earlier Kondratiev cycles (better named 'surges') is of lower quality than economic data available today, the value of such a perception is that it focuses attention on the interrelationships of technology. Indeed, it is regrettable that the input/output tables of modern economies have not been redrawn to take account of the changing structure of technology, as Leontieff originally intended. Leontieff himself is acutely aware of the importance and pervasiveness of information technology as a determinant of future economic evolution.

I have argued that a detailed prognostication of the impact on employment is extremely difficult, though all the indications are that the consequences will be immense. It is now necessary to consider future developments in technology. Little attention has yet been paid by any analyst to the tremendous multiplier that information technology and the rapid advances in computer power have given to both applied research and development, and indeed basic science. A veritable cornucopia of new products and new processes can be expected, and from one perspective there need be no shortage in the long run of goods and services.

Even this perspective, though, serves only to highlight the transitional problems of adjustment in the short and medium term. Indeed, the *cri de coeur* at the heart of this chapter is that the traditional methods of allocating economic activity and its rewards, both between and within economies, are likely to produce stresses which may be unacceptable at many levels. It is perhaps ironic that the UK, which benefited for so long from being the first nation-state to industrialise, may be the first to be faced with the unpalatable consequences of adaptation to the new economic age. For many reasons, not least that the social resilience of the UK has historically been high, it seems important that these issues be discussed, even though some of the more extreme predictions prove unduly doom-laden. If this chapter has contributed to that end by describing potential problems, then so be it. The most useful role of forecasting is to suggest potential problems ahead, so that strategies to avoid these problems can be found.

REFERENCES

Barron, I. and Curnow, R. *The Future with Microelectronics*, p. 191, (Frances Pinter/Open University Press, 1979).

Porat, M. U. *The Information Economy*, vol. 1, (Centre for Interdisciplinary Research, Stanford).

Barron and Curnow, *op. cit.*, p. 191.

Kondratiev, M. D. 'The Major Economic Cycles', in *Vorprosy Conjunctkury*, vol. 1 (1925).

5 A Trade Union View

BARRIE SHERMAN

Trade union membership in Britain is not a minority taste – as is sometimes suggested. There are now over 12 million members in unions affiliated to the Trades Union Congress representing well over 60 per cent of the eligible work-force. The number of members is rising and when this is viewed against the background of considerable media hostility, it represents a remarkable achievement. When it is then also looked at in terms of the new membership this feeling is reinforced. White-collar workers in both the public and private sectors have provided the bulk of this increase with the more traditional craft- and industry-oriented unions losing ground. Many of these new members have no previous union experience and often no social reinforcement of family ties with unions. Clearly a need has been perceived by employees, and one of the major components of this need is protection against insecurity. Change, especially change out of one's control, breeds insecurity and instability and over the past few years there has been a considerable amount of such change.

Both the local authority and health service structures have changed and thus employment and promotion patterns within them, whilst overall the units increased in size; firms have merged and been taken over with the same sort of results; the UK entered the Common Market and government's policies changed along with international trade patterns and responses. Throughout this last decade the unemployment trend has been upwards, especially for young people, partly because of these structural changes, partly because our industrial base has declined dramatically and partly because of contractionary government economic policies.

We are now standing on the threshold of an even larger change. A technologically based society where microelectronic applications will change productive, distributive, and service processes across the whole range of British industry and commerce, as well as change and amend many products. It is difficult to think of a single sector which cannot and will not be affected other than some personal services. It is equally difficult to envisage the economic, social, and political changes that may well be necessary to successfully encompass these changes. It is not something that the UK has had to face before in the twentieth century – in this respect it is a unique change. From the more distant past, the late eighteenth, early nineteenth centuries, the omens are not too good; we did not cope overall with those circumstances and the expectations prevailing today make it far harder to succeed.

Western (indeed the Comecon countries which are not immune from these developments) societies are now based on rising expectations. These include not only the expectancy of more and better material goods but also services such as health and education, better family security and not the least important, political advancement. Whilst on the face of it this new technology would enable us to achieve all the aims simultaneously it will happen neither overnight, nor of itself. The leads and lags involved will provide immense problems in the short and medium terms whilst the final objective could well be overtaken by short-term political events. In other words it is a *dynamic problem* and thus not one that can be satisfactorily resolved using comparative static analysis – the path that is chosen (or stumbled upon) will determine the destination.

If we compare our standard of living with that of our predecessors at the beginning of the nineteenth century, romanticism apart, today it is better in almost every respect. Material goods, transport, entertainment, nutrition, health, longevity, anything we now take for granted would have seemed like science fiction to our forebears. Yet all these improvements have come about through technological changes, some small, some large. Steam, electricity, internal combustion engines, oil, artificial fibres, antibiotics, etc. When we look back it is tempting to see the path from then until now as a smooth continuum – a stroll through bluebell-covered technological lanes – and then predict that the future will be no more difficult or problematical. *Yet nothing could be further from the truth. It was no smooth pathway that our forebears trod – it was misery; anguish and political, social, and economic upheaval all the way.*

The steam engine was developed to draw water out of the mines. When it became the motive power behind the textile, coal, steel, woodworking and engineering industries it stimulated not only the first industrial revolution but also the growth of new cities and towns. This period saw the growth of Britain as the world's premier industrialised nation with cheap food and raw materials from the Empire, which also acted as a market for our manufactured goods. Yet was it a good time to be alive and working? No. The golden age was more like dross for the majority of the people.

Epidemics, sickness and a life-expectancy of 40 years were the most people could hope for. Bad housing, 6-day weeks, 12-hour days, children in the mines and weaving sheds – hardly something to brag about. Workers moved off the land into the towns and cities and production grew, but not smoothly. There were slumps and over-production, people were laid off and starved. Even with the low expectations of the time there was massive discontent. The Luddites (who opposed deskilling, not job loss, and were right) took their direct actions; the Chartists grew in numbers and Britain trundled nearer on the brink of a revolution than ever before. There had to be a political Reform Act in 1832. It was a turbulent, worrying period.

The century ended as Britain was dragging itself out of its deepest-ever industrial slump (either before or since) which had lasted for over 20 years. There had had to be another Reform Act to extend the suffrage and there was also a new technology – the electric motor. Both this and the steam engine had two key factors in common; they were used primarily on the supply side of the economy, this is, they acted as a method of producing things rather than changing the things themselves. No-one went out and bought a steam engine, nor an electric motor. Both were incorporated into capital machinery. They both acted as powerful stimulants to productivity and efficiency. The second key point is that they were ubiquitous, all productive processes used them, they became *the* motive powers. There were few services and service industries, few offices and clerical type jobs, small government and local government establishments. Overwhelmingly manufacturing industry was the primary factor of this period.

Nearly all of the twentieth century technological changes to date can be divided into two categories. They either produced a new consumer product, for example the mass-produced motor car or antibiotics, or they had a limited impact on the supply side of the economy. The limitation was that only one or two industries changed at any one time. Thus it was posssible to box off the consequent unemployment or training effects and we all had the luxury of time on our side to enable the economic multiplier effects to work through the economy and reap the benefits of the increased productivity and lower real prices of the goods. Even the large main frame computer introduction, much heralded, had limited effects – though not as small as some commentators assert.

The main-frames were greeted with cries of doom and gloom, especially in the USA. It was thought that they might replace workers to a far greater extent than in fact actually happened but in hindsight it is quite clear that this would actually be the case. The main-frame was cumbersome and non-flexible, and few people, especially those in senior managerial positions, knew what they could do or how they could do it. Enterprises bought or leased the machines only to find that their internal information systems were not providing the type of information needed in the form in which it was needed. Not only were employees in aggregate not laid off – more had to be hired so as to obtain and process this information. The second factor was that the systems often went wrong; the whole 1960s period is littered with computer horror stories. Each time this happened the more unscrupulous data-processing manager would plead a shortage of staff and as no-one in a more senior position was able to contradict this because of zero knowledge of the matter, departmental staffs grew. The bubble burst in the late1960s, since which time department size has been reduced, information systems installed and natural wastage is being allowed to take its course. It would seem from recent research in Cambridge that there is an 11- or 12-year lag between the introduction of the computer and consequent direct job loss although it has been argued (Jenkins and Sherman, 1977) that there has been substantial indirect job loss through mergers etc., which could not have been even envisaged without computerised techniques.

The problem with the early 1960s gloom-mongering was that now we do have a real problem it appears that we are crying wolf, and there is a tendency in some circles to deride the arguments. Whilst it is easy to argue this point of view it is almost certainly wrong for two reasons. The first is that even the little boy who cried wolf was confronted by one in the end. The second is that microelectronics are not, despite initial appearances, merely computers. They are cheapish, small and flexible. They can be used everywhere, not just in a specially prepared room; they can be used as components as well as in capital equipment, and in short they are nothing like computers as we know and understand them.

This chapter is about the trade union view of micro-electronics and this, in itself, presupposes that there is something important to have a view about. Several things disturb the trade union movement on this subject. The first is that we will not use this new technology and if we do it will be too late. The second is that if past industrial relations and management mistakes are repeated we shall provoke industrial disputes. Finally that unless pre-emptive action is taken the technology will be introduced on the 'backs of the workers'; there will be job loss, there will be skill changes and dislocation and all these costs will be borne by workers.

For years the annual Trades Union Congress has been inundated with resolutions about the need for more domestic investment so as to re-equip British industry and make it competitive once again. This has been reinforced by officers and stewards of unions at company and enterprise level; trade unions know only too well the British de-industrialisation problem. There are many causes for it. Inflation, lack of consistent effective demand, strange foreign exchange policies allied to balance of payments problems, spare capacity and a very risk-avertive management system are amongst them. Unions have asked for the protection of import penetration ceilings, but been refused; asked for consistent demand policies but been refused, and asked for protective legislation to bring Britain onto the levels of its European counterparts – and it is being dismantled. Whilst such pleas go unheeded there is a growing feeling in the movement that the benefits of new investment and new technologies will go to one set of people but the costs fall on union members.

Unions are in business to protect and represent their members individually and collectively. Historically this has been accomplished by a series of short-term reactions rather than longer-term forward planning. This is partially due to the election, representation and communication systems of unions and the manner in which their structures have evolved, and partially due to the fact that unions have been shut out of the decision-making processes at national levels as well as at company levels. In circumstances such as these there is no alternative but to react to given events. Members of unions, whilst never welcoming conflicts (they are unpleasant, money is often lost and relationships soured), do not expect their union to agree to everything a management proposes. The interests of management are very rarely if ever the same as the bulk of the work-force no matter whether it is a publicly or privately owned organisation. Senior management will wish to sales-, profit- or service-maximise and cost-minimise (given that they have other constraints) whilst trade unions will wish to maximise the amount of employment and salaries of their members – this must be a conflict position. What makes this so important is the history of industrial relations in Britain. If either management or unions make a statement the protagonist will invariably place the worst possible construction upon it (if alternative interpretations are possible). The media, too, often use this to provide sensational stories and so often exacerbate a situation or even provoke a dispute. From the union point of view they feel they have justification for their attitudes. Companies who have said they were doing well laid people off; companies which claimed penury when salary negotiations loomed made record profits, and in the public sector the same sort of things have occurred. It is against this sort of background that the introduction of new technologies must be viewed.

It does not mean that no co-operation will be possible, and nor does it mean that conflicts are inevitable. However it does mean that attitudes, especially about the managerial prerogative and information transmission, must change if trouble is to be avoided. Should the same attitudes prevail it is evident that there will be defensive actions by unions – this will lead to the cry of Luddites by the media and management, yet those who write and those who manage would themselves be loth to be made redundant, especially when jobs are more difficult to come by. Nor indeed would many of them wish to involuntarily move to a new job, have to retrain in a totally new skill or see themselves passed over for salary increases or promotion. Yet these are things they will be attacking others for refusing to accept. There needs to be sympathy and changes in attitudes if difficulties are not to 'choke-off' the technology boom.

Microelectronics can reduce the number of available jobs and change the skills needed to do those jobs. It does so over the entire economy at roughly the same time. There will be fewer places to hide, few jobs totally unaffected. I shall deal with job changes first.

Some traditional (or quasi-traditional) jobs will become redundant. *The Times* Newspapers dispute has thrown the problem of skilled printers into sharp relief; not only would there be fewer jobs but the technology changes the remaining jobs out of all recognition. Computer-aided design is changing the skills of the draughtsman as well as the engineer; word processors are changing the skills of typists, and numerically controlled machines are changing the skills of engineering craftsmen. One can run through other trades such as woodworking, ceramics, footwear manufacture, and clothing and see that various skills are no longer needed. Generally, though not always, the changes are of a de-skilling nature rather than skill-enhancing. However other changes occur. Previously set promotion patterns are disrupted as new skills, often brought in from outside the enterprise, suddenly become important. Employees have to retrain (and sometimes move). Job evaluation schemes become irrelevant. In other words a previously stable in-work environment suddenly becomes unstable.

This all creates fear and worry. Fear of losing one's job and fear of not being able to cope with new skills or machines; fear of not being able to cope with the retraining and fear of losing one's relative position in a hierarchy. Employees realise that their promotion prospects have been blighted and no longer try as hard – why should they? These changes, which all happened when the main-frame computers were introduced into the insurance industry, can have a major impact on the industrial relations climate. Either the employees rushed into trade unions for protection or those who were already members moved into more militant postures. On another level some employees take a pride in their jobs; only some because so many jobs are so boring, repetitive, and easy to do that pride is not a word that can readily be associated with them. The jobs that do attract it are typically those for which people have had to get some sort of qualifications or serve an apprenticeship. To make a skilled engineer into a machine minder or a professional engineer into a keyboard operator and expect them to show as much enthusiasm as previously is asking just too much. With this sort of phenomenon superimposed on a more general shift, including changes in managerial responsibilities and functions, the stable background for a smooth transition is just not present.

Microelectronics can lose jobs in several basic ways. The main one is on the direct production side and here there are two distinct ways in which loss can occur. The first is through robotics and other automated equipment. This is the science fiction, good media-presentation, stuff, although unfortunately for film-makers the robots look more like giraffes designed by Emmett than 3CPO. Robots, which can undoubtedly replace people, suffer from the present disadvantages of being heavy, expensive and prone to breakdowns. They are perhaps a technology of 10 years' time rather than the next 2 or 3 years although for matters such as welding they are quite adequate. Cheap computing power provides a plant, especially a small one, with a method of becoming very efficient. Production planning, materials handling and control, critical path analysis all become very feasible at very low cost. Big, fast main-frames are not needed, small second-hand computers which can chug away all night if necessary are quite adequate providing the programmes are right. Such a change, dependent as it is on relative cheapness, will mean either employees will be redundant or output can be increased using the same number of employees. Once one or two firms in a market start to use these techniques they become more efficient and take a larger market share. This results either in competitive firms going out of business or emulating the original firm – either way, jobs are lost unless the total market expands at least as fast as the rise in productivity.

Whilst both of these methods could cost a lot of jobs, especially in the small-firm sector which has up to now been more labour-intensive in overall UK terms, they will cause the least problems. This is because less than 10 per cent of the work-force now work directly on production. Of far greater significance in employment terms is the use of microelectronics as a component.

Wherever there is an electromechanical device or yards of electrical wiring it is possible to replace them with integrated circuitry. This generally has the advantage of being smaller, cheaper, more reliable and is more easily serviced so that there is an incentive to use them. When the Japanese TV industry doubled the number of the integrated circuits in their colour TV sets they also reduced by half the numbers of other components – transistors, resistors and diodes, etc. The total number of people involved in the manufacture and assembly of the sets was halved whilst output went up over 30 per cent. It proved to be a better, cheaper to run and very easily serviced set which virtually destroyed the British TV manufacturing industry. If one looks closely at these sort of ratios, that is a drastic reduction in the number and size of components, it becomes clear that they will be the levels of the future. It means that an awful lot of jobs are at risk.

If there are fewer components, fewer (and different) people will be needed to manufacture the components, fewer to assemble them, instal them, invoice them, bill them, store them, warehouse them and transport them. The production process becomes truncated and thus fewer supervisors and managers will be needed. In short, it hits at the bulk of employment, the non-direct workers in manufacturing industry.

The third way of losing jobs is in the service sector. Microelectronics have been said to stimulate the information revolution and the bulk of office work is concerned with information. The word processor is the key to office changes. In itself it is an efficient machine which should increase productivity in various ways but affect only typists and those who need documents or letters typed. The potential of the word processor, however, starts to be fulfilled when it is linked to others through the (soon to be completed) all-electronic telecommunications system. When this happens word processors will talk to word processors – it will be a form of electronic mail. When voice-recognition activation arrives this compounds the problem (assuming that we do not all have to adopt Japanese accents). It was thought that word processors would be clustered around minicomputers or even main-frames, but although this is still possible the processors themselves now have enough spare computing capacity to do different tasks, *viz.* cash flow estimates, etc.

In terms of employment numbers and changes in hierarchies these developments are very significant. 'Clerical and related' employment accounts for nearly one job in five over the whole economy and this does not include general management and administration. The recent decline in manufacturing industry employment has been matched by an increase in service and other industry employments. The finance, shipping and commercial industries, the local and central government offices are all stacked full of administrative employees who spend their time shuffling bits of paper – for the most part usefully. The problem arises when these bits of paper are no longer needed; filing, administration and a host of back-up jobs disappear. The German company Siemens, which manufactures some of this equipment, has estimated that 40 per cent (2 million) West German clerical jobs will not be needed by 1985. The French government report 'L' information de la Societé' (Rapport à M. le President de la République by Sumon Nora (Inspecteur General des Finances) et Alain Mine) estimates that 30 per cent of French finance industry jobs will go over roughly the same period.

These three basic ways of losing jobs do not include the retail and wholesale trade problems, draughtsmen, printers and journalists, etc. Wherever a job exists that needs little or no personal discretion, imagination or man-management abilities, it is possible to replace it by a machine. Herein lies the major problem. The reason why precedents this century are of little help is that it is so widespread. It happens all over this and other industrialised societies at roughly the same time.

It is probably the worst of all possible times to introduce this technology. The growth of world trade has been slowing down from well before the 1973 oil crisis, which only reinforced the process. The immediate post-World War II growth rates are unlikely to be repeated. The present energy crisis will itself act as a further brake on growth so that the chances of high export-led growth for the UK correspondingly diminish. When this is taken in conjunction with the fact that nearly all new investment over the past decade has been to increase efficiency and productivity rather than to cope with expanding markets (OECD McCracken Report), it is clear that our ability to supply is out-stripping our ability to demand. This is precisely what happened between 1874 and 1895.

Another disquieting thought is that of the so-called 'super-competitors'. These are the newly developed or highly developing countries – Taiwan, South Korea, Brazil, and Singapore are among them. They are using cheap and 'captive' labour together with very sophisticated technology to compete in the industrial growth sector.

Consumer electronics, shipbuilding, steel and cars are amongst their range. The days when a developing country was expected to plough through all the traditional heavy industries before reaching an industrialised status are long gone. The new technologies, although sophisticated, are relatively cheap in capital terms and often very simple to use; developing countries can afford to buy them in, can use them productively and can leap-frog the traditional path. This clearly puts extra pressure on the industrialised countries both in terms of competing in world markets and in the need to adopt and adapt to the new microelectronic technologies.

There is a school of thought which states that although the theory behind job loss may be correct offsetting factors will make the problem vanish. There are three basic views of this. One is that economic multiplier effects will work, a second that new products will emerge and a third is that public expenditure can be increased. All three hide a multitude of political and economic problems.

In the UK we have particular problems with multipliers in a dynamic sense. Although there may be some effect with diminishing real prices of manufactured goods, the pricing system is not such for these to be easily obtainable in the UK. If there is increasing unemployment it leads to diminished aggregate demand and what is worse, those who are the most likely to be unemployed are those who typically spend (rather than save) the highest proportion of their earnings. On the profits side they can only be re-invested (thus making the problem worse) or distributed and these latter tend to go to financial institutions – which do not consume. We are thus most unlikely to see a vast stimulus to employment from this area.

No doubt there will be some new consumer products – although at present no-one can actually think of many examples. They must be new goods for new markets. New goods for old markets merely create unemployment – as in the watch industry when digital watches were introduced, or the desk calculator industry when electronic calculators replaced them. Most of the present microelectronic applications are to amend and improve existing products, for example automobile applications, electronic timers on cookers, etc. Whilst these will be sought after and may stimulate an earlier replacement cycle, they will not add to overall aggregate demand.

The final point is the most political. There is no doubt that there are needs to be met not only in the UK but in the whole world. Health care, education, social care for the mentally and physically handicapped and care of the aged cannot be said to be perfect in any country, let alone the starvation and malnutrition endemic in many parts of the world. To meet these needs generally, means a person-to-person contact, and many jobs can be created. Leaving aside the training and retraining problems they also all need public expenditure and this raises the question of how the monies get to the Exchequer in the first place and from whom or what. As this becomes a question of the distribution of income and wealth it is intensely and fundamentally political. There is another option – that of work-sharing and changing our patterns of work and attitudes to work.

When people become involuntarily unemployed they lose in three distinct ways. Their health suffers. Recent research by H. Brenner (1976) has shown a statistically relevant relationship between unemployment and high morbidity and mortality rates both in the USA and Britain. People also lose financially. Unless an unemployed person is very badly paid, unemployment leads to a quite drastic reduction in family living standards and expectations. The third way is psychological. People have an ingrained work ethic and tend to identify with their jobs whilst society (or parts of it at least) reinforce this by suggesting that the unemployed are 'scroungers' or 'spongers on the state' and are somehow morally defective. We thus push people into believing that they need to work.

This is a dubious premise. The work ethic really emerged with the first industrial revolution when all the establishment organs, from the church to magistrates, promoted it. If cleanliness was next to godliness then work was only a short head away in third place. However, as people did not actually enjoy their work over-much a dichotomy arose (and still exists) where they felt they had to do something that they disliked – a syndrone best exemplified by the British attitude to spinach or senna-pod tea. People certainly need to be needed. Whether this *must* be achieved by working for other people is quite a different matter. I believe not, although clearly it will take many generations of hard counter-propaganda to eliminate the ethic, so deeply is it now ingrained.

This can be seen quite simply in the trade union movement's reaction to work. The right to work is not only a slogan of movement, it is a tenet of faith. The young right-to-work marchers appear regularly at the TUC Congress, and each individual union is under pressure to minimise job loss by its membership. This is not surprising given the disadvantages that accrue to the unemployed. What follows from this is that if the only way to use the new technologies is to make people redundant, then the stigmata – financial and otherwise – that attach themselves to the unemployed must be removed. In essence we must look at unemployment as an investment for the future and we do not penalise investments, indeed, we encourage them with all sorts of allowances, grants and tax concessions. *Why should people be treated more disadvantageously than capital?* Whilst this is then partially a union function it is also a major responsibility of government.

A shorter working week is, however, a totally different matter. It is in itself without any unemployment pressure, a laudable collective bargaining objective. The working week has fallen over time from 6 days to 5, from 60 hours to 40, whilst paid holidays are basically a post-war matter. A shorter working week is thus good in its own right as well as useful in spreading employment around. But three points have to be made. The first is the simple one on Britain's competitive position. We live in a low-pay economy with fewer holidays than our competitors. Whilst it is true that a massive shortening of the working week with no reduction in earnings would have a deleterious competitive effect, there is a considerable margin within which improvements can be made without significant damage. Clearly a large change should be on an international basis, be it an international union negotiation with a multinational company, or various international body (like the EEC) initiatives.

The other points are on totally different matters. The first conerns the now well-known European Trades Union Confederation demand for a 35-hour week. This would not significantly dent present unemployment levels let alone what is probably going to happen. On an hourly basis it would have to be under 20 hours per week to solve the problems of the 1990s; but even this is not the major point. If people do not like work, they like travelling to and from work even less, as this involves rush hours, traffic jams and stress and general strain of crowded buses, a trains and tube. It would be better all the way around and avoid the inevitable unsocial shift systems associated with shortening hours per day if days per week were shortened instead. The trade union movement should (and will) demand a 4- or 3-day working week as the main plank in a work-sharing plan. Indeed this point was made by the proposers of the 35-hour week resolution at the 1979 TUC Congress.

One other matter is clear. It is quite lunatic to run a society where people work hard all of their lives and have their only large chunk of leisure at a time when they are least equipped to use it – that is at retirement. Surely it is not beyond the wit of man to arrange work so that this can be avoided: to have long holidays; to have regular sabbatical leaves; to have part-time retirement. There are a whole range of proposals which could be implemented to achieve the end result and for the first time there is a technology that could enable it to be done. *The sum total of goods and services will be produced using fewer resources including manpower*. In trade union terms this is the crucial point. It releases people from the necessity of drudgery and yet maintains living standards.

This then becomes less 'work-sharing' and more 'less work'. It also implies a more equitable distribution of wealth. Better working conditions and less of it overall are, howerer, ends. It is a dynamic problem and the 'means', the path which is taken, is the second crucial point.

All politicians of whatever persuasion ultimately believe in virtue. A better life for all people and all their families, rising incomes, more freedom, these are the stocks in trade of politicians whether amateur or professional. It is the method by which one reaches this moving utopia that differs so much. So it has been with the rudimentary political debate, rudimentary because the official lines of both parties have been jingoistic and naive. The debate, if it can be described as such, has centred around the advantages of the technology and completely ignored the costs. Trade unions which by definition are practical, were not and are not satisfied with wooly obfuscations – there must be practical policies. If the British government continues to pretend that everything in the garden is lovely then they cannot take the pre-emptive actions needed to smooth the transition. If they cannot take these actions then the reactions are likely to be formidable. If we are confronted with policies which are devoted to merely letting the market work then in all senses we are back into the early nineteenth century. Can any rational person believe that, short of banning trade unions and applying the law by force, present-day workers will tolerate the conditions of that period, or even of the 1930s even if these are upgraded into modern relative terms. Of course the answer must be NO.

Whilst this is yet another fundamental political question the trade union movement is wedded to planning in ordinary circumstances and in response to short-term problems and constraints. This technology alters the basic framework and can scarcely be described as ordinary. The possible, indeed probable, dislocations are immense, ranging through unemployment – especially amongst the young and women – training and educational needs, housing and other mobility requirements, fear and a host of information, proper job placement provisions, etc. To expect the market to do all this is, to the trade unions, a nonsense in a modern industrial state where markets are, to say the least, imperfect both nationally and internationally. Furthermore such a system would, in the opinion of the trade unions, expressed at Congress after Congress, only serve to increase the inequities in income and wealth distribution.

If the first thing that we have to do is use the new technologies and this involves some unemployment, then surely the job of any government is to make this acceptable and to arrange the environment so that there is the maximum chance of success. Yet this does not appear to be happening; indeed, over the past few years we have moved in precisely the wrong direction. By cutting public services, by widening the financial difference between those employed and those unemployed, it is far harder for trade unions to avoid resistance even if they wanted to. Even if there were rational negotiations and the unions did their job well, using natural wastage and voluntary early retirement, the burden of unemployment would fall on the young and school-leavers. This is the major frightening prospect.

The British education system teaches children to read, write, and do their sums, because if they do they will get a job, not because they are good things in their own right. But what happens when this contract breaks down and the jobs are not there, haven't these young people a right to be angry. In Britain this anger tends to manifest itself in hooliganism and vandalism but in other countries the anger is more politically motivated into terrorism. Could that happen here? Could it be imported here from EEC countries with similar problems? These are questions which are better not answered – rather, they should not be necessary to ask. School-leavers tend to be unrepresented. Too young to vote, not members of unions, no political parties other than those on the fringes both left and right to encourage them, they are in a power and influence vacuum. Government, unions, and management, all bear a responsibility in this matter and must hammer out with young people coherent policies and actions at all levels.

Unions have two areas of action. There is now a national role through the NEDO tripartite system and the various quasi-government bodies on which they are represented, as well as direct meetings with Ministers of the Crown. Collective bargaining is, however, the traditional part that unions play and equally traditionally this has been confined to terms and conditions of service. Whilst this will still be very important, in recent years the movement has become more ambitious and wishes to become part of the decision-making process. Technology will hone these ambitions. I shall concentrate on the negotiation level as it is here that the crunch, if any, will come.

The Norwegians have had technology agreements for many years. These enable stewards to meet and discuss with employers the purchase, siting, and repercussions of new equipment or systems well in advance of even the decision to purchase. This idea is now starting to catch on in Britain and will form part of the two-pronged attack – shorter working weeks being the second. The book *Collapse of Work* (Jenkins and Sherman, 1979) and the TUC document on technology both float the idea though neither in detail. It is worth spending some time on such an agreement.

The idea is for the trade unions to be able to influence, if not control, the new technologies and both their rate of introduction and their implementation. One thing that is almost certain is the likelihood of unions banding together to negotiate these agreements as they get together on health and safety matters and pensions negotiations. In many respects, in many sites in Britain this will be an excellent thing and in all probability will stimulate union mergers eventually. The agreements will be formal and the procedures negotiated, hopefully in calm surroundings rather than the frenetic circumstances surrounding industrial disputes. Nevertheless it is quite clear that disputes will occur in some places.

From a trade union point of view the most important clause will be the first – a *status quo* arrangement – without which the agreement will not be worth the paper it is written on. This, in essence, means that nothing will be changed or introduced until procedure has been exhausted. It can be qualified of course, as in the ACTT/IBA Agreement, where this lasts for a specified time period; it can be limited in other ways too, but it must exist. The remainder of the agreement can be divided into two separate parts: regular and early information and consultation, and the actual bargaining about practical matters.

The information is vitally needed and as explained earlier is long overdue in Britain. It must not be of the normal '*fait accompli*' type, and should be early enough to pre-date the actual decision-making process and realistic enough to stand the test of time. This not only requires a change in the attitudes of trade unionists evaluating given information, but requires a far greater change in the managerial ranks in their guarding of information on dubious grounds. If other countries can divulge information to unions without commercial secrecy being impaired, without industrial espionage running riot, and without the collapse of the managerial prerogative, why not Britain? The method of information exchange and consultation is probably best carried out in the formal setting of 'technology conferences', held at least four times a year between management and unions. They must be held at the appropriate levels; that is to say, the levels of decision-making, bargaining, and implementation, and it is possible in large multi-plant, multi-product and multinational companies that these will not coincide.

Trade unions will need to know what new technologies *may* be introduced and what new systems will emerge as a result. If there are choices of equipment which will impinge on the working environment then the unions will wish to be in on the purchasing process. Manpower forecasts, training and retraining needs, recruitment policies, skill shortages, health and safety implications and perhaps job security matters, will all properly belong to this forum. Finally, work methods will need to be negotiated as well as time off for the union representatives to attend courses, etc. This last point is important in that whilst the union representatives must not just get involved in academic discussions they must be able to understand and query the statements made to them.

There are both direct and indirect effects of technological change. A radical change in shop floor equipment and procedures will affect those workers directly but will also affect clinical, administrative, and managerial workers in indirect ways. The same applies to changes in components and products and often with the procedure reversed. A radical re-vamping of an internal (or even external) communications system will often affect shop floor and other departmental systems. It is for this reason that there will have to be multi-union committees and a high degree of sophistication attached to them.

However these sort of meetings are nothing like the be-all and end-all of the matter. New technologies impinge on various aspects of collective bargaining and many of these will need to be settled before it can be said to have been satisfactorily introduced. Salary levels are the first and most obvious matter. New skills and new responsibilities often require higher remuneration whilst conversely other people will want compensation for the blighting of their promotion prospects or for a new and thoroughly disagreeable job. Whilst these have to be negotiated through the normal bargaining procedures there is also the problem that nearly all existing differentials and thus job evaluation schemes will be upset. Indeed on the latter point, new skills, etc., will make a nonsense of existing schemes and many will be far beyond repair by the existing review committees and will need to be totally reworked.

Productivity schemes and payments are another matter. It is evident that productivity, especially in offices, will rise and it is equally clear that unions will want a share of this increase. Redundant employees are often those who have made productivity increases possible, and it is highly probable that unions will firstly create a special sort of membership for this sort of person, and secondly have them receive productivity payments in addition to those for the remaining work-force. Whilst opinion varies amongst not only unions but also different groups in the same union as to the desirability of productivity arrangements, it is very probable that they will loom large over the next decade. It is incidentally interesting to look at value-added schemes. When companies used their own internal components and made them from blanks, etc., value-added was high, but the bringing in of ready-made integrated circuits changed this. It is likely that the value-added schemes will be challenged and die the death.

Hours, weeks, and other working arrangements will need to be negotiated as will shifts, overtime, and other premia. Health and safety matters will be of paramount importance and, at present, there is considerable disquiet on the effects of visual display units and also daisy-wheel printers on eyesight and hearing. Pension scheme arrangements may have to change. Both preservation and transferability arrangements penalise people changing jobs. As the need to change increases and as the mobility of people is put at a premium so those existing arrangements will be seen to be unfair. A 'club arrangement' or some contract scheme is desirable.

For unions this all involves a major change. The shop stewards or representatives will have to exercise more discretion and power. Communication systems will have to be improved (the new technologies will come in handy here) and education provisions enhanced. It would seem that the basic Donovan Commission's recommendation of basing industrial relations where the power is (on the shop floor) is slowly to become a reality.

No sector of the work-force will be immune. Teachers, doctors, skilled and unskilled industrial workers, finance or retail staffs, Civil Servants and footballers, all will be affected in various degrees. The trade union movement is now very broad, it encompasses all those and more. It has members right up to board level in the private sector and very senior Civil Servants in the public sector, and all have one thing in common: a perceived need to be represented collectively. In bargaining and in central representational terms this makes the trade union movement stronger than it has ever been. Yet something still is missing.

Decisions on political and industrial matters are still taken by a small number of people. In industry and commerce the multinational corporation has stood conventional business economics on its head. What does one of these companies try to optimise in either any one country or globally? In politics, governments and their advisors tend to look at the next election and no further, and Stock Exchange system reinforces this short-term approach in what is erroneously called a capital market. *But the problems caused by microelectronics are longer-*

term. Herein lies an important dilema. Bodies which are adequate for short-term decision-making will have to attempt to plan long-term and there must be some degree of training interest in this change.

Two other matters are of great concern. Risk capital is almost non-existent in Britain where our capital is tied up unproductively in insurance companies, pension funds, unit trusts and building societies to a great extent. Yet we will need risk capital which to date has had to be provided by the government through the NEB and other agencies. British management for various reasons tend to avoid risk and often avoid clever people. Whilst this obtains we will not get an internationally competitive industrial base despite incentives, despite planning, and despite organisational changes.

Attitudes will need to alter. If we are confronted with many women and many young people involuntarily unemployed the stimulus for attitude changes is there. If we wish to substitute leisure for work, if we wish to reduce the amount of work, the stimulus is there. If we wish to have a better and more secure life for our children and grandchildren the stimulus is there. The trouble is that these things will not happen of their own accord – we, people and their representatives, will be responsible for seeing that these things happen. There is of course the stimulus for other things to happen. For the rich nations to grow wealthier at the expense of the poorer. For repression and totalitarianism and for the rich to literally grind the poor. We will all have to make choices, not least the trade union movement, but until unions have the power which goes with their strength they can only have a peripheral, and almost by definition, negative influence. It is in none of our interests for this to remain the case.

To get the desired ends one has to adapt the right means. Mistakes, however, will choke off development of the technology. We face the most crucial decade of this century and whilst it would be nice to say we do so with confidence, this is hardly true. If policies, both short-term and what pass for long-term, do not change, if it is not possible to negotiate reasonably with employers – and both will need changes in attitude – we will have little chance. So it must be done, and unions must see it done.

REFERENCES

Brennes, H. Report to the VS Congress Joint Economic Committee on economic effects of unemployment.

Jenkins, C. and Sherman, B. *Computers and the Unions* (Longman, 1977).

Jenkins, C. and Sherman, B. *Collapse of Work* (Methuen, 1979).

6 Automation and Industrial Robots

JOSEPH J. HUNTER

1. INTRODUCTION

Digital computers have been around for some time and their use in manufacturing is not a novelty, numerical control (NC), computer-aided design (CAD) and computer-aided manufacturing (CAM) are well enough known terms in this context, although many companies might regard them as not being very relevant to their operations. It might be asked what has changed with the advent of the computer on a chip – the microprocessor. The answer lies simply in the enormous decrease in cost and increase in power of electronic devices over the last two decades which has made their use potentially all-pervasive. Every manufacturer must now examine his product to see if it can be improved by incorporating electronics, and also his market to see if new products can be introduced.

Correspondingly as every manufacturer is also a user of someone else's improved or new product, it follows that changes are likely which will affect his own organisation and production methods. This encompasses all that has formerly been understood by NC, CAD and CAM, and much else besides.

In the following sections a brief look will be taken at microcomputers, the new equipment which incorporates them, and their uses. An area which will be looked at in more detail will be numerical control of machine tools, industrial robots and their application to the automated factory.

2. MICROCOMPUTERS

The factors in the evolution of digital computers were first the realisation that information could be put into digital form and second that, in this form, it could easily be transmitted, processed, and stored. Of some importance is the fact that major parts of the equipment required to carry out these functions could be mass-produced using assemblies consisting of large numbers of similar elementary circuits. A further major advance, the microelectronics era, has resulted from a technology with which these circuits can be printed on to silicon wafers. The growth in semiconductor technology, which led to the ability to put a computer onto a single chip of silicon, has been very rapid. For example, in 1965 the semiconductor manufacturers could put 30 components on one chip. This component density doubled every year, reaching 30,000 components in 1975 with a corresponding thousand-fold reduction in cost per component. In 1978 the 64K-bit RAM was produced with 135,000 components on the chip. The 1 million-component chip is now in sight. It is of interest to note that eventual limits to this growth can be seen, due to the physical difficulties of manufacturing devices with extremely small dimensions, but within these limits it appears likely that single-chip microcomputers can be made which are powerful enough for most tasks.

The circuits which make up a computer can be partitioned into a few main units. The processor which interprets the program and carries out arithmetic and logical operations on the data, the memory in which the program and the data can be stored, and input and output circuits through which information is received and transmitted.

The initial advances with large-scale integration (LSI) were made with memory circuits; the 'microprocessor' followed later. There are two main kinds of semiconductor memory: random access memory (RAM) which is alterable and also volatile, i.e. loses its contents if the power is switched off; and read-only memory (ROM) which is unalterable after the initial information is stored in it. There are also intermediate memory types such as electrically programmable read-only memory (EPROM) which is alterable, although not easily, and so is often used in the development phase of equipment design.

Although these microelectronic circuits have been developed for use in computers, their low cost and power make them universally applicable wherever measurement, processing, storage, display or transmission of information is required, or control of a process or machine. Many of the applications found in industry turn out to be even more suitable ones than general-purpose computing. For example in a machine controller, the set of instructions which determines how the machine operates is generally fixed and can be programmed in ROM, thus providing security of operation. Data which only exists during the machine operation can be held in RAM. There is often no need for any other kind of storage either on-line – e.g. magnetic tape or discs – or off-line – e.g. paper tape, or for keyboards or elaborate displays. It is these 'peripheral' devices which can comprise the major part of the cost of a general-purpose computer. For example it is estimated that computer costs can be broken down into 1 per cent for the central processor unit, 5 per cent for the peripheral circuits, 20 per cent for the memory and 74 per cent for the peripheral devices.

It should be appreciated that the use of microelectronic circuits of this type in equipment such as controllers, sequencers, monitors, etc., constitutes a major change of practice with implications both in design and maintenance. For the original equipment designer, the end resuit of the design process is now a set of operating instructions in the form of a program which can be built into the equipment. The electronic circuit behind the front panel might be exactly the same for many different applications. In contrast, the end result of the design process would previously have been a list of circuit elements and a wiring diagram showing how these were to be connected. In addition the maintenance engineer would have been furnished with a circuit description showing the voltages and waveforms to be expected at different points in the equipment. In order to find faults in microcomputer-based equipment above the trivial level of power supply failures, etc., it may be necessary to examine the sequence of instructions which appear as digital signals on the terminals of the circuits. Special items of equipment, such as logic state analysers, are becoming available for this purpose. Oncc the signals are obtained, however, and have been checked against the correct machine code instructions, it may still be difficult to determine where a fault originates. To overcome this problem, special test sequences may be built in at the design stage of the equipment which can be used to isolate faulty chips. Maintenance work at this level is a job for a technician, and suitable personnel are likely to be in short supply. On the brighter side it might be noted that the incidence of faults with LSI circuits is very much lower. In addition it is possible to build in extensive self-checking facilities and fault diagnostic routines all of which considerably speed up fault-finding and repair. Changes to the content of technical college and university courses in electronic engineering which reflect these demands are being made now. Existing personnel will require training courses.

The equipment manufacturer who is thinking of using 'micro-processors' has a number of options to choose from.

The most basic level is to start from the microelectronic chips. This approach entails the highest development costs but gives the lowest production costs. Typical uses would be in high-volume applications, e.g. TV games, motor vehicles, washing machines. Chip costs might be in the range from £5 to £50.

The next level up is to start with the chips pre-assembled onto a circuit board in the form of a 'single-board computer' and a large variety of ancillary boards. This has many advantages for the manufacturer with a lower volume of production, e.g. up to 10,000 of a particular equipment. Single-board computer costs range from £100 to £1000.

For both of these approaches, program development requires to be carried out on special development systems and the cost both of this equipment and the program development effort can be substantial. As an illustration, it might take a day to produce five lines of correctly working program with an average line length of eight bytes. The operating program for a not very complicated application could be 1000 bytes in size, i.e. about 6 man-months of effort.

The remaining possibility for the equipment manufacturer is to incorporate in his design a proprietary microcomputer-based equipment configured and preprogrammed for his application area. In this case only a relatively small effort is required to complete the operating program for the specific application under consideration. There is already a large variety of microcomputer-based equipment of this type. Costs range from £500 to over £5000. Application areas might be for machine and process controllers, particularly where these are expensive and with a low volume of production.

Users, as well as equipment manufacturers, are also likely to be interested at this level. Some of the application areas in manufacturing are discussed in the following sections.

2.1. SINGLE-BOARD COMPUTERS

Single-board computers are complete modular assemblies of a microprocessor chip, ROM or PROM program memory chips, RAM data memory chips, input and output chips and a clock oscillator on a plug-in board. The memory can usually be expanded, or other functions added to the system, by using additional plug-in boards. Four-bit and 8-bit single-board computers are available both from independent specialist companies and from semiconductor manufacturers; 16-bit single-board computers are available mainly from minicomputer manufacturers.

Microprocessor development systems are required for programming the 4- and 8-bit single-board computers. These are available from board manufacturers such as Intel, Motorola, etc., or from electronic test equipment manufacturers such as Tektronix, General Radio, Hewlett Packard, and Philips. These latter companies produce 'universal' systems that can be equipped for different types of microprocessor Costs for these disc-based systems range from £3000 to £25,000. The cheaper 'development' units consisting of a microcomputer with a keyboard and display are suitable for small jobs and for educational purposes.

Development systems for the 16-bit microcomputers are composed of one of the minicomputers produced by the same firm which in many cases is already available. This has the advantage that high-level programming languages can be used and a considerable amount of existing software is accessible. For this reason the LSI-11 single-board computer, manufactured by the Digital Equipment Corporation, has been used extensively in universities and research establishments.

A noteworthy development has been the introduction of standard 'bus' systems for the single-board computer. In a farsighted move one of the large microelectronic chip manufacturers, Intel, has put its system, the MULTIBUS, into the public domain and published full details of the board connectors, signals, and voltage levels in order to encourage other manufacturers to use it. Over 100 MULTIBUS-compatible boards are available from dozens of manufacturers. These include support boards such as memory, communications (digital and serial I/O), analogue input/out (I/O), high-speed maths processors, peripheral control, and direct memory access control.

In a recent development Prolog and Mostek have combined to design the STDBUS. This is meant for smaller boards than the MULTIBUS. There are a number of other bus systems.

3. APPLICATIONS

There are a number of areas common to all businesses where microcomputers might be used, e.g. order processing, purchasing, wages, purchasing and stock control etc. Areas in manufacturing which are of particular interest are listed below.

(a) *Design* of equipment, layout of printed circuit boards, production of drawings, shape nesting.
(b) *Scheduling and process planning* of work to machines, planned maintenance, on-line workshop control.

(c) *Machine and process control*, e.g. numerical control of metal-cutting and forming machines and industrial robots, sequence control of machines and conveyors, boiler and furnace control, batch and continuous process control.
(d) *Machine monitoring* for productivity, performance maintenance and safety.
(e) *Inspection*, e.g. in-process and between-process measurement and inspection of parts to detect faulty operations and reduce the production of scrap.
(f) *Plant and factory communications* to transmit production schedules and other information to personnel and to permit data collection at the point of manufacture.
(g) *Mechanical performance testing*, e.g. hydraulic machinery, engines and engine parts.
(h) *Electrical performance testing* of components including integrated circuits, subassemblies, and complete systems.
(i) *Plant and factory facilities monitoring* for consumption and condition of supplies, e.g. air, water, fuel and power, etc. Also site security and safety.

4. MICROCOMPUTER-BASED EQUIPMENT

In all of the application areas noted in the preceding section, microcomputer-based equipment is already available. Examples are found in programmable logical controllers (PLC), measurement and control systems, data monitors and loggers, production monitoring systems, security and alarm monitoring systems.

These new microcomputer-based equipments offer greatly improved facilities over previous equipment designs based on discrete components. As it is difficult to cover this wide range of equipment in one chapter, three areas will be discussed in the following sections that are particularly relevant to the theme of the automated factory.

The first is the desk-top computer which is the most general-purpose of the microcomputer-based equipment. Some examples will be given to illustrate the wide usefulness of desk-top computers in design and test applications.

In the second area, the benefits conferred by the distributed computing power of microcomputers will be examined in the case of numerical machine tool control units.

Finally the enhancements to the capabilities of industrial robots that are possible with microcomputer control units are discussed, and the combination of industrial robots with NC machine tools in the computer-integrated factory.

4.1. DESK-TOP COMPUTERS

Machines in this category are also being termed programmable calculators, instrumentation controllers, and lately 'personal' computer systems.

The characteristic being advertised by their manufacturers is that they are 'friendly'. By this they mean that they are easy to use and require no specialist staff either for programming or maintenance. They also operate within the range of environment conditions normally found in offices, factories and homes so that special premises are not required.

The main difference between desk-top computers and minicomputers is that the desk-top machines are complete and ready to use with input/output devices and local storage. The operating system and language interpreter are built-in hardware. Programmes, which are directly interpreted at execution time, can be entered and edited from a keyboard. Magnetic cards or tape cassettes are provided for programme storage and re-use. Programme running time is longer than with minicomputers and direct memory is usually smaller. These are the main limitations.

Desk-top computers configured for engineering and scientific applications are widely available. The first company in the field was Wang, closely followed by Hewlett Packard who are probably now the main manufacturer and also offer the most extensive range of peripherals. The Tektronix type 4051 and peripherals is suited to the computer-aided design (CAD) area. Personal computers like the Commodore PET offer similar facilities at low cost. Both IBM and Olivetti have desk top machines useful for computer-aided design purposes.

Languages

The earlier programmable calculators, and presently the bottom end of the Hewlett Packard range, are programmed in a similar manner to pocket calculators. Each keystroke corresponds to an operation such as multiply, divide, sin, cos or to a digit. Stores have to be assigned by the user. Hewlett Packard uses a stack and reverse Polish notation logic to carry out calculations.

The most common language on the more sophisticated desk-top machines is BASIC, often with considerable enhancement. The IBM

5100 is notable for offering APL as an option. A feature of some machines is the optional plug-in ROM available that greatly increase the power of the machine in specific areas, e.g. graphics or sophisticated input/output operations.

In applications where a keyboard input, a display and relatively frequent changes of program are required, a desk-top computer is likely to be a good choice. These conditions are met in CAD, in mechanical and electrical performance testing and analysis of experimental data and in automatic metrology. The manufacturers provide some often-used programs with the equipment, e.g. mathematical and statistics libraries, others are available at a modest charge and there are also users' clubs where programs can be interchanged. In addition there are independent firms who sell specialised programs, plug-ins and peripherals.

One of the following examples illustrates the use of a desk-top computer for CAD; in another it has been incorporated into a measuring machine by a manufacturer and finally a similar development by a user.

High-performance mixed-flow machine design system (HPMF)
This is a suite of programs, developed at the National Engineering Laboratory (NEL) and used on an HP model 9830 desk-top computer with an HP model 9862 plotter for the complete hydraulic layout of mixed-flow fans.
A series of programmes perform the following functions:

(1) rotor and inlet casing design;
(2) axial stator design;
(3) annular diffuser design;
(4) scaled plots of the entire casing;
(5) rotor blade detail design;
(6) stator blade detail design;
(7) rotor and stator blade stacking and the plotting of orthographic, perspective and stereo views of any part or all of the machine from any viewing point;
(8) the calculation of blade surface co-ordinates for a specified blade thickness superimposed on the blade camber lines;
(9) the printing of any or all of the co-ordinate and other tables on the line printer for direct inclusion in the design report or on the plotter for inserting into engineering drawings.

A 'link' programme combines the above programme into a complete system and allows data to be stored in, and retrived from, a common data bank. Permanent storage of the data bank allows a design to be continued over a series of sessions.

Metrology and quality control

The Carl Zeiss Company use (Newman, 1976) an HP 9825 to control their Zeiss UMM 500 Universal Measuring Machine and to process the measurement data. An automatic calibration routine using a sphere or cube standard is carried out before the start of each measurement to determine the ball centre and diameter of the probe being used. The UMM 500 can be combined with a controlled indexing table (Zeiss RTO5) for rotary measurements.

The Ford Motor Company at Indianapolis (Thomas and Mock, 1978) are using the HP 9815 for process control gauging, machine adjustment determination for set up operators lot sampling. It is also used as a controller for a Multiple Dimensional Inspection and Analytical Device (UMDIAD). This consists of a group of standard electronic gauging units interfaced to an HP 9815. Sixteen separate and independent measurements can be made at a rate of six per second.

4.2 MACHINE TOOL NUMERICAL CONTROL

The numerical control (NC) of machine tools was first envisaged as a means of solving difficult problems such as machining aerofoils, airframe sections and other complex shapes. Soon after the first NC machines were developed it became apparent that the 'part programmer' who translated the drawings of a part into the input data for the computer required assistance to reduce the calculating effort involved. This led to the development of computer programs such as APT (Automatically Programmed Tool). These programs required what until recently were considered large and expensive computers; consequently the computing work was organised in such a way that it could be concentrated onto a single large computer. A diagram of the steps involved in going from the drawing to the finished part is shown in Figure 6.1. It can be seen that there may be up to three departments involved, i.e. design, data processing and production. The APT and post-processor program runs, although shown separately for clarity, would be carried out sequentially on the same computer. In a chain like this there is obviously room for errors being made. There are diagnostic checks at the computer output stage for various faults but the final test is at the machine tool. Errors at the workpiece are not uncommonly due to the mistakes in part-programming so that corrections have to go around

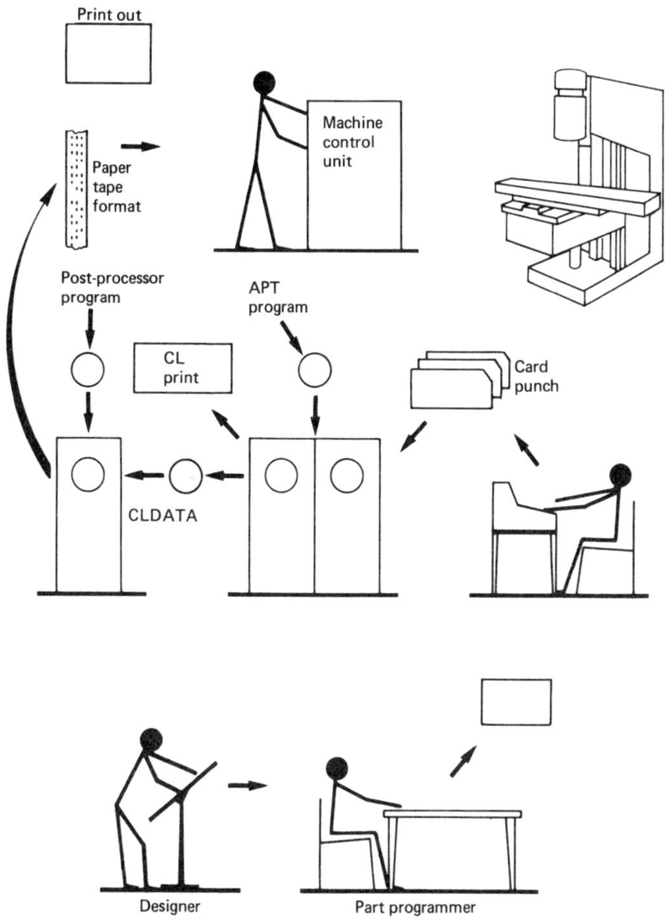

FIGURE 6.1 A 'conventional' system

the whole loop. This may take days or even weeks, which is one of the main problems with numerical control as it has evolved up to now.

Although the numerical control of machine tools was introduced to deal with complex shapes, it was soon appreciated that it could profitably be applied for simpler purposes, where little processing was required, such as drilling repetitive patterns of holes. This quickly led to a proliferation of machines equipped with numerical control systems with either 'manual data input' (MDI) where the input instructions could be set on keyboards, switches or plug boards, or with coded instructions.

These were two distinct directions in which numerical control had developed but in both, the machine control unit, being based on 'hard-wired' electronics, was limited in capability. It had been appreciated for a long time that all the processing and control functions required in the machine control unit could be carried out by a digital computer. With the decrease in cost which became apparent with the advent of the 'minicomputer' the first 'computer numerical control' (CNC) machine control units became available. With the advent of microcomputer machine control units, hard-wired electronic control units have been almost completely eclipsed. Some of the greatly improved features and facilities which are available with microcomputer numerical control units will be listed below, but it is worth noting that the cause of divergence in numerical control equipment practice has disappeared. This is a consequence of the 'distributed' computing power which is now affordable with the microcomputer.

Among the features that can be found on the new microcomputer numerical control units are:

(a) *Post-processor* functions built into the machine control units. This trend has been taken to its logical conclusion in the Vega IIIG which accepts a CLDATA input.

(b) *Standard interface* – most new machine control units are provided with an RS232C standard interface. This allows connection to a variety of input and output devices and also computers.

(c) *Manual data input* is provided as an optional input mode. The input instructions for some units have the same form as the part programming language which is used for off-line programming. In others labelled buttons, which correspond to machine functions, are used to enter the program.

(d) *Programming aids* such as visual display of programmed shapes, menu selection of program steps or program prompts, mirror-image capability, 'canned cycles' for repetition of certain operations.

(e) *Local storage and editing* of part programs in the machine control units. A number of part programs can be stored (equivalent to several hundred feet of paper tape).
(f) *Fault-finding with diagnostic display*, automatic monitoring of interlocks; oil pressure, motor temperature, etc., with a built-in fault diagnosis program which can indicate the part that requires attention, greatly simplify maintenance and fault-finding requirements.
(g) *Tool nose radius and tool offset compensation, tool data files* are features that ease the job of the manual programmer.
(h) *Automatic calibration* of slide position against a reference such as a laser can be stored in a programmable read-only memory (PROM).
(i) *Load, temperature, and backlash compensation* can also be stored to allow on-line correction. With these features and provision for automatic calibration machine tool accuracy might be improved by an order of magnitude at little extra cost.
(j) *Adaptive control* of machining speeds based on torque or power monitoring, or on chip-making detection, can increase the machine work rate.
(k) *Workpiece, tool handling and inspection devices* can be connected as peripherals to the machine tool allowing for automatic operation with automatic resetting of tools to compensate for wear.

Many of the new units are composed of several microcomputers. This makes it easier to design, implement, or add on the features described above. No machine control unit has all of these features, to our knowledge, although this would be possible as they are not very expensive to implement. Taken together they suggest that much of the technology for automatic, unattended operation of machines in the computer-integrated factory already exists. One of the problems in fitting them together is that of standards; for example, making the machines and computers talk the same language.

4.3. INDUSTRIAL ROBOTS

There has been some controversy over just what is an industrial robot. A definition which clarifies this has recently been proposed by the ISO Committee TC97/SC8.

> A machine formed by a mechanism including several degrees of freedom, often having the appearance of one or several arms ending in

> a wrist capable of holding a tool or a work piece or an inspection device. In particular its control unit must use a memorizing device, and sometimes it can use sensing and adaptation appliances taking into account environment and circumstances. These multipurpose machines are generally designed to carry out a repetitive function and can be adapted to other functions.

This definition excludes manually controlled manipulators, or tele-operators, and simple actuator devices which cannot be reprogrammed. However it is useful in examining the scope of the industrial robot to look at some uses of these simpler devices. For example there are in use programmed manipulators which have two or three axes of movement powered by pneumatic or hydraulic cylinders with their extension set by mechanical limit stops and having a fixed sequence of actions. In Japan in 1976 it was estimated that there was about 17,000 of these devices being used for unloading plastic injection moulding machines, 1700 for pouring metal and unloading die-casting machines and 6000 for loading and unloading presses.

Fixed-sequence and movement manipulators like this are quite satisfactory for mass production but not for smaller batch production where it is necessary to be able to quickly reprogramme for different tasks. Their main benefit is that they enable a higher production rate to be attained by the costly machines with which they are associated than is possible by human operators.

Arc welding is one of the newer application areas in which there has been a recent increase in interest with sales of over 100 robot welding systems being reported by ESAB and the entry into the field of BOC Ltd in association with Hall Automation.

Initially the electronic control unit comprised a large part of the cost of these robots. In the same way as for numerical control, microcomputer control units are decreasing the cost and greatly increasing the capabilities of industrial robots. A great deal of interest is now being taken in more advanced applications. For example (Abraham *et al.*, 1977), General Motors have a project for a programmable universal machine for assembly (PUMA) and Westinghouse have an adaptable programmable assembly system (APAS). A robot being used in these projects is the newly introduced Unimate 500 which has a multi-microprocessor design, a high-level programming language and a price of £18,000 for one robot reducing to £12,000 for larger numbers. The Unimate 500 has reach and load capacity similar to a human so that it can take over assembly line tasks and be interchangeable with human operatives. Cincinnati Milacron produce a larger computer-controlled robot, the 6CH, which is being installed by General Motors in one of the first industrial applications of interactive vision. In this a video camera scans a conveyor belt carrying a mixture of castings. Different castings are recognised and sorted by the robot into boxes or other conveyors. The 6CH robot is also being used for drilling and de-burring holes in aircraft sections and then riveting them. Other uses are work piece handling and arc welding. Other robots with microcomputer control and advanced capabilities include Olivetti's SIGMA assembly robot, Asea's type Irb-6 and Irb-60 robots which are used for polishing, grinding, welding and workpiece handling and Hitachi's 'Mr Arcs' welding robot and their paint spraying robot.

The first industrial robot, the Unimate, introduced in 1962, had a means for easy programming. In appearance the Unimate resembles a small tank turret which can rotate and tilt and has a retractable gun barrel. On the end of the barrel is mounted a 'wrist' carrying a gripper which has another two axes of rotation. The gripper, which can hold either a tool or a workpiece, can be moved through a series of positions under manual control in a 'teach mode' by operating switches. Any desired position can be recorded and the sequence of movements between taught positions can then be repeated indefinitely. The speed at which the sequence of movements is carried out can be independently adjusted. The path between the taught points is not controlled in the 'point to point' mode of operation. A later model allows straight-line movement between points at a controlled velocity in a 'continuous path' mode. In the later models of the Unimate, the 'memorising' device is a plated wire memory and a memory capacity of up to 1024 points is available. Taught programmes can be recorded from the memory into a tape cassette, or an external computer and quickly re-loaded when the same job has to be repeated. The original application area for which the Unimate was designed was in die casting where productivity increases of 35 per cent have been obtained for an expenditure on the robot installation in the order of 10 per cent of the cost of the large die casting machine. As an illustration of the flexibility of use of the industrial robot, it turned out that the main application area for Unimates, with about 1000 in use worldwide, has been in spot-welding car bodies.

Some application areas such as spray painting, coating, arc welding, and grinding require the ability to repeat movements in a continuous-path mode. The Norwegian company Trallfa produced the first industrial robot for these uses and has sold over 800 for painting and coating. This is also the main application area for the only UK manufacturer, Hall Automation, whose latest robot models have a microcomputer-based control unit. The operation of Occupational Safety and Health Acts and reluctance of people to work in these areas is causing a move towards greater usage of industrial robots.

The number of industrial robots in use today is minute compared with the human working population. In part this is due to the robots' limited capabilities. For example robots have almost no ability to deal with variability in the work or its positioning. Work must be jigged or fixtured and many apparently unskilled tasks which contain some element of visual inspection or workpiece manoeuvering cannot be done. But, as with NC machine control units, this position is changing with the advent of microcomputer-based control units and the development of robot senses, such as vision.

A German study (Warnecke and Schraft, 1977) indicates the potential market for present-day industrial robots in a range of metal-forming and machining applications is about 2 per cent of the workplaces. With the improvements which are forecast in industrial robots, they expect the potential market to increase to about 35 per cent of the workplaces by 1985. The potential sales market in West Germany for industrial robots could then amount to about £500 million per year. The Japanese market value estimate (Yomemoto and Shiimo, 1977) for 1985 is similar except that it is for predicted sales rather than potential and includes fixed-sequence manipulators. The number of robots involved in these predictions is in the tens of thousands per year; still small compared with the number of jobs in manufacturing industry.

The other limiting factor in robot growth is cost, both of robots and for changes to working practice. Here real improvements in living standards have a dual effect of reducing the cost barrier and also increasing people's expectations with regard to their work environment. In both Germany and Japan, Government programmes aimed at promoting robot use and manufacture have emphasised the theme of 'humanising' the workplace and improving the quality of working life.

Within the next decade there is unlikely to be much opposition to increased robot usage because of unemployment. In the longer term, however, the effect will be cumulative. The combination of industrial robots, numerical control of machine tools, CAD, manufacture and overall plant control leads to the concept of the computer-integrated factory. This is particularly relevant to about 40 per cent of all manufacturing industry which uses batch production methods and is labour-intensive. The forecast is that this area will follow the mass production and process industries and become largely automated.

Employment in manufacturing industry would decrease from about 25 per cent of the total work-force to under 5 per cent by the year 2000.

5. CONCLUSION

The low cost and power of the microcomputer makes it attractive for very many applications in industry. The change to 'distributed' computing and processing power will affect nearly every sector of industry. The greatest effect is likely to be felt in the small-batch production area which comprises about 40 per cent of industrial production. The combination of numerical controlled machines with robot work-handlers, integrated together on a computer network, is providing some of the tools required for the automation of small-batch manufacture. Large gains are possible in increasing the utilisation of machine tools, and even greater gains in decreasing the amount of work in progress. There have been other earlier attempts at achieving the computer-integrated factory, e.g. the East German Prisma and Rota factories which have suffered from over-centralisation. The opportunities now afforded by microcomputers are much more extensive and more flexible than has hitherto been visualised as computer aided design and manufacture. Because of the capital investment in plant, and because the changes in organisation will be more difficult, it is likely that changes will take place over a much longer period than those which are likely in the office side of industry. The automation of small-batch production is being promoted in the UK by the ASP project which is supported by the Department of Industry's Mechanical Engineering and Machine Tools Requirement Board.

ACKNOWLEDGEMENTS

Our thanks to the Director of the National Engineering Laboratory for permission to publish, and to colleague Mr T Stirling for supplying details of his application.

REFERENCES

Newman, J. H. 'Precision measurement'. *HP Keyboard.* vol 8, no. 3.

Thomas, G. P. and Mock, S. G. 'HP 9815 in quality control'. *HP Keyboard,* 1978, no. 4.

Abraham, R. G., Stewart, R. J. and Shum, L. Y. *State of the Art in Programmable Assembly Systems* (International Fluidic Services Ltd, Bedford, May 1977).

Warnecke, H. J. and Schraft, R. D. Applications of industrial robots – premises and volumes. ECE Working Party for Automation, Seminar on Industrial Robots and Programmable Controllers, Copenhagen, Denmark, September 1977.

Yomemoto, K. and Shiimo, K. 'Present state and future outlook of technology and market for industrial robots in Japan'. *Proc. 7th ISIR*, pp. 37–48 (Tokyo, Japan, October 1977).

7 The Impact on Production Management

PAUL F. BESTWICK and A. E. OWEN

GENERAL INTRODUCTION

Around the world, leading manufacturers are firmly established in a trend to reduce or eliminate man from the shop floor and production lines. The major reason is the realisation that man contributes nothing to the production process that cannot be done better or more efficiently by automation.

There are many managerial arguments for automation; these range from labour-saving capital equipment in countries where the cost of labour is high, through enforced usage because of the reluctance of the local populace to work in the machine shops and production lines, and most importantly, because of the need for savings in work in progress, speed and reliability of throughput and consistency of product (Hollingum, 1979b).

There is a common belief that direct labour is the primary contributor to the factory cost of a product. However an investigation into manufacturing operations in the UK (New, 1976) found that the overall median cost breakdown as a percentage of full factory cost is:

bought-out materials and components	51 per cent
direct labour	18 per cent
overheads	31 per cent

From this it can be seen that reduction in costs pertaining to materials is three times more beneficial than reduction in direct labour costs. It should also be noted that there are synergistic rewards from reduction of materials costs, in that the introduction of a given material-handling system can result in a reduction in the work in progress and, at the same time, mean a reduction in the requirement for direct labour.

That work in progress is responsible for a large proportion of incurred factory costs pertaining to materials, can be seen from reports (Dudley, 1970, 1975; New, 1976) that cite median throughput efficiencies of 21–30 per cent in batch production facilities. This means that for more than 70 per cent of its manufacturing lead time, a component is involved in avoidable non-productive activities. Management in this area of responsibility of manufacturing industry, should therefore concentrate its efforts on the introduction of automation that will:

(1) enable a reduction in material and material-handling related costs;
(2) improve the management of the conversion process, thereby reducing the non-productive element of the conversion process.

It is suggested that the application of microprocessor units (MPU), in manufacturing industry, should be concentrated on the management of the conversion process, rather than on the technological applications to the facilities within this process. It is to these areas that the MPU has almost limitless potential. It is the cheapness, reliability and lack of any special environmental need, in the form of temperature and humidity controls, which allow the MPU to be used by line management in a production environment.

Thus, this chapter is concerned with the application of MPUs to the production manager's task. Within these tasks problems can be classified as programmable or non-programmable. Non-programmable problems cannot be 'routinised' and are those which must be dealt with using personal skills, experience, judgement and enterprise. Programmable problems can be delegated to subordinates; their structure is known, and established techniques can be used to assist in the analysis. It is to these areas of decision that the MPU has its largest potential use.

Initially the requirements for an integrated system structure are presented. This is followed by examples of MPU computerisation of systems within various industries. Finally the 'logical' conclusion of the application of MPUs in industry – 'the computer-automated factory – its development and potential realisation is discussed.

INTEGRATED SYSTEM STRUCTURE

INTRODUCTION

In any company it is 'top management' that has ultimate responsibility, since it is they who, through example and direction, determine how a company functions, and whether it survives. In order to make strategic decisions it is necessary to be able to evaluate the company, such that the company's 'personality' does not cloud the issues.

It has been stated that, whilst division of the whole is destructive, it is necessary to segment in order to understand (Lockyer, 1975). Segmentation ensures that it is possible to get a basic understanding of the overall system structure of the manufacturing company, and of the relationships between that structure and overall productivity. There also needs to be a fundamental logistic strategy for deciding what should be automated, when and how. This concept is directly opposed to the long-standing belief held by management, that all plant operations other than receiving, manufacture, and shipping are of little consequence. It seems that management has difficulty in comprehending that work-flow in a company is managed, not by a set of independent systems, such as purchasing, inventory or bill of sale, nor by several sets of integrated systems such as materials requirement planning (MRP), accounting, shop-floor control, but by one logical overall system structure (Appleton, 1977; Cox, 1977). Adoption of this concept will not only lead to a system structure that promotes productivity, reduces costs and overheads, improves quality and stimulates on time production; it will also lead to automation using the database concept.

WHAT NEEDS CONTROLLING?

The amount of data generated in a modern manufacturing corporation is vast and can be of prime importance to decision-makers. Because these corporations have become so diversified and sophisticated, it is paramount to schedule, plan, arbitrate, expedite and co-ordinate the activities of man and machines. Such management implies flows of information through the plant: upwards for reporting status to supervisors, downwards for the issuance of shop orders and laterally for synchronising related operations.

In order that management can control these activities, answers to the following questions would be useful for making both short-term decisions and long-term plans:

(1) What shop orders are delayed, overdue or assigned special priorities?
(2) What parts are needed and where can they be obtained?
(3) What alternative work centres could be assigned to the job?

(4) What are the deviations from plans or schedules and what corrective actions are possible?
(5) What are the current lead times, delays and scrap rates on the various machines?
(6) What men, machines, tools, parts, and supplies are available?

To obtain answers, it is not only necessary to acquire data, but to maintain current dynamic files, which reflect inventory, labour force, work in progress, shop orders and sales, as well as other factory status conditions.

At the lowest common denominator, the database, all manufacturing companies look pretty much the same, the major variations occurring because of the different reporting requirements. For example, although their decision problems are different, managers in both process and job shop operations use basically the same data. In short, from a database standpoint, manufacturing companies look different primarily because management information demands are different, not because the companies use different data.

FUNDAMENTAL PIECES

Using the database technology, the system structure is divided into three fundamental pieces (rather than '*n*' pieces); one for each special computer application (Appleton, 1977).

The first piece is called the *database* and is simply a collection of data elements organised with some kind of structure. These elements are the vocabulary of the business, the words that the managers use to ask questions, understand answers, and make decisions.

Since the database reflects the system structure of the business, rather than the organisational or financial structure, it accordingly reflects the customer order, or part orientation, of the system in its pure or hybrid form, i.e. in the pure job shop, the data would be structured around the customer's order, whereas in the process shop it would be geared to parts.

The second critical piece in the system structure is what could be called the *input control system*, which is generally geared to the organisational structure and is hence independent of the database.

The third piece of the system structure is the *outgoing control system* which is not similar to either the first or second pieces, because the output demands are always changing whilst the database and input control system are relatively stable.

Hence the output control system must be designed as a dynamic system and responsiveness to all sorts of enquiries is mandatory. The flexibility of this latter piece of the automated system structure is critical to the survivability of the whole.

STRUCTURAL LEVELS

In a typical manufacturing plant, data is created and used in one or more of four basic structural levels (Kochhar, 1977).

The material flow level

This refers to the transformation of raw materials into finished goods. It is the only level at which physical movement of goods takes place, either in the form of the production of goods or the receipt of bought-out items, and requires a great mass of detailed information.

The process control level

This is concerned with supervision, quality control and process unit control. In consequence only a small amount of information is required and, in general, the time duration for which the information is kept after creation is short. Only in the case of abnormal operating conditions is it necessary to store information for future analysis.

The production control level

This is concerned with the hour-to-hour management and optimal utilisation of the production facilities. It is therefore interested in scheduling, cost control, inventory control and maintenance. It requires information of lesser detail than at the process control level, but needs to store significant information on the data base.

The management control level

This has two sub-levels:

(1) *Strategic*, which is concerned with overall policy and requires brief summaries of past operations and information relating to external factors that influence the management of the company, for instance market demand forecasts, current economic conditions, competitors, raw material availability, costs and social pressures.
(2) *Tactical*, which is concerned with putting the policy into practice, e.g. making production plans meet demand and making efficient use

of resources. Therefore more detailed information is required regarding internal operations than at the process control level.

The role of management is to plan and control operations in the company; it is therefore of the utmost importance that it be given the tools with which to carry this out. The overall efficiency of the manufacturing system is dependent upon the efficiency of production at the material flow level as well as the timeliness of information flows at the other levels. Because, with the exception of the management control level, the majority of the information is dynamic, it is important that the computer control system is equally dynamic.

COMPANY MATURITY

The totality to which system control can be obtained is dependent upon the 'maturity' of the company with respect to the 'pure process' shop in which the highest levels of productivity can be attained. The evolutionary process can be visualised as occurring in four stages (Appleton, 1977).

Job management

This is the first stage. Here basic controls over purchasing, engineering, manufacture, and other production activities are exercised by the customer's order: hence cost and inventory controls are weak. The system structure is flexible and geared to reschedule management; in consequence, overheads are high and redundancy is low.

Standard production activities are treated by the system as being exceptions to the normal procedure. Although in the overall concept of system structure this stage is the least efficient, productivity can be improved by a reactive management, with strong controls and using data of the current overall order status and impact analysis for schedule and priority changes.

Product definition

The second stage, product definition, is where the company seeks higher productivity, hence standardisation of product lines or portions of the more standard products takes place. This involves the development of part controls and a refinement and restructuring of the database. The objective now is to define the products not as new designs, but as variations of the current design of the basic parts manufactured by the company.

One of the important steps at this stage is the concept of group technology (GT). Effective use of this form of classification will provide the versatility and control over product variation needed to make the psychological switch to variation management in all production phases including engineering, manufacture accounting and marketing.

Part management
Arrival at this stage is through the adoption of systems such as materials requirement planning (MRP), the dominant controls of which are oriented towards parts and processes rather than customers' orders, because of the requirement for consistency, predictability, discipline and most importantly, accurate timely feedback.

Computer-aided design and computer-aided manufacture (CAD/CAM)
These are indicators of the arrival at the stage of highest productivity. At this stage there is full integration between the automated administrative control structure and the CAD/CAM technology. This is not to say that CAD/CAM cannot be used at other stages, only that it would not be fully integrated and hence not fully efficient.

The four stages are not independent of one another, since each must be established upon its predecessor and, in consequence, poor development of one stage will hinder effective development of its successor. Not all companies proceed through all of the stages and migration from one stage to another can take several years. However, it should be the objective of every manufacturer to move as far as possible in his own environment towards the ideal of the pure process shop. This objective is not unrealistic since everybody likes to do better.

The real issues are, at what stage should the company in a given market really be? And how effectively is it operating at its present stage? Only management can answer these questions.

COMPUTERISATION OF SYSTEMS

INTRODUCTION

Since their introduction into manufacturing industry in the early 1950s, computers have been used for a wide range of activities. Whilst, in general, these have been the automation of existing manual procedures, there has been an ever-growing requirement for computers to take over the chore functions of the manager – the programmable problems. The availability and capability of the modern high-speed computer, especially the microcomputer, has meant that the planning and control of the day-to-day running of the factory can be accomplished by the manager, often in real time (Kochhar and Parnaby, 1977, Kochhar, 1978).

When discussing computer systems, one hears such terms as main-frame or maxicomputer, minicomputer, distributed or network or web minicomputer and microcomputer. In recent years it has become increasingly difficult to produce clear definitions for each type since, for instance, a modern microcomputer can be as powerful as a main-frame computer of a few years ago. In very general terms, the four types of computer systems can be defined in terms of their role, rather than their size or capacity:

(1) *Main-frame computer*: large, general-purpose, multi-user high time utilisation, requiring an interface of highly trained specialist staff.
(2) *Minicomputer*: restricted to a clearly defined range of activities and a common user purpose application which may be fairly broad. Has to be highly effective when used by ordinary users with only limited assistance from specialists. May not have high time utilisation.
(3) *Distributed minicomputer*: a number of minicomputers linked together as an alternative to a main-frame computer. The low cost of the individual units means that the costs are spread across various departments, and it becomes economic to have a spare standby unit to cover the group. Because the individual minicomputer is used for a restricted range of local processing, it reacts faster for a given operation than would a main-frame computer. Also, since the system is kept simple and is not sophisticated, it is less likely to fail.
(4) *Microcomputer*: usually dedicated for a single specialised application and often has very low time utilisation.

Computer systems operate in one of two time modes, dependent upon the type of computer and the information being generated. Since the computer tends to be of one mode or the other, it is the 'urgency' or complexity of the problem that determines the computer used.

Urgency is directly related to the time-span of action for a given activity in the manufacturing industry. Therefore, if the time-span is long, it is probable that a batch mode computer will be used, whereas if the time-span is short then it is imperative that an on-line real-time (OLRT) computer is utilised. The importance and implications of the time-span of action of activities within the manufacturing environment is emphasised in a recent article (Bestwick and Lockyer, 1979).

BATCH LOADED MODE

Computers were developed initially for commercial data-processing in companies that could realize significant benefits from the investment. Because of the huge investment, it was necessary that these main-frame computers be operated in the batch loaded mode, such that maximum usage and consequently maximum efficiency of the facility be realised.

Whilst the main-frame computer is ideal for the commercial functions of a company, and for the evaluation of trends based on historical data, such that forecasts can be generated for management strategic decision-making, it is not suitable for the evaluation and solution of immediate problems.

These immediate problems come about because the structure of a typical manufacturing organisation consists of a number of departments carrying out related, but different, tasks on different time-scales. This means that at any one time there are many different areas of responsibility within the company, all wanting data in a style that is pertinent to their needs.

Because shop-floor plans can go amiss due to random occurrences such as breakdown of machines, absenteeism amongst workers and the non-availability of material, it is necessary that the 'time-span of action' for a given order be available as and when required. Therefore, due to the dynamics inherent in the day-to-day running of the factory, it is essential that the computer system that is used to control the factory be of a nature that is at least as dynamic as the interactions themselves.

ON-LINE REAL-TIME MODE (OLRT)

The requirement for a dynamic computer system has resulted in the development of the minicomputer and especially the microcomputers. These computers have real-time operating capability which greatly enhances managerial control.

OLRT systems have the extremely important characteristic of

allowing man–machine interaction since, because of the continual communication between user and computer, the user is able to use his own adaptive thought processes as he pursues information, before responding to the output from the computer (Kochhar and Parnaby, 1977).

The creation and maintenance of up-to-date files, reflecting the status of a plant at any instant, permit tracking of the flow of materials, monitoring work in progress, allocation of labour resources and the scheduling of short-term production changes (Cox, 1977). This maintenance of files can now be accomplished directly by the relevant person through the use of voice input terminals (Myers, 1978). Also, intelligent terminals that have been 'ruggedised', for the use and abuse of production personnel in the plant environment, are readily available (Cox, 1977; Large, 1979 (a) and (b)).

OLRT computer systems do not have to operate in the batch loaded mode for their utilization to be economically viable, since whilst they have a very reduced efficiency, when compared to main-frame computers, provided that the increased productivity generated through their usage pays for the computer capital and running costs, and still results in increased profits, then the efficiency of the OLRT system is irrelevant.

A POSSIBLE CONSEQUENCE OF A SMALL COMPUTER INSTALLATION

The proliferation of small computers within a company is responsible for the gradual demise of the authority of the powerful central computer, cosseted in its controlled environment. It is likewise responsible for the loss in status of the data processing (DP) departments in corporations (Ward, 1979; Wood, 1977).

Because these small computers, which are controlled through an inexpensive microprocessor, are usually linked into the main-frame computer, a 'Gulliver and Lilliput' situation arises. This situation arises when the corporation decides to replace or update its main-frame computer. Because of the low capital outlay, the small computers are purchased rather than leased and, in consequence, once installed remain installed. Hence, if the newly leased main-frame computer or the new software cannot communicate with the smaller computers, it is the main-frame computer that is crippled, not the minicomputer or microcomputers. There is, therefore, a situation whereby a small computer in, say manufacturing, restricts the freedom of the DP department to choose the central computer system.

The importance of centralised managerial control, such that chaos does not ensue, and that maximum benefit be derived by the company as a whole from its computerised systems, cannot be overstated.

COMPUTERISATION IN THE SMALL COMPANY

Minicomputers and microcomputers are inexpensive, partly because of the big market and the consequent use of high-volume production techniques. The market for the small computer is growing at a much faster pace than that of the main-frame computers; also, because developments in the field of smaller computer systems are rapid and since production runs tend to be short, improvements in technology can be passed on to the customer very quickly. Because one can buy 10 to 200 minicomputers for the price of one main-frame computer, even the smallest company can afford to install a minicomputer or microcomputer system (Kochhar and Parnaby, 1977).

In fact, in survival terms alone, they cannot afford not to 'go computer', because profits are made and lost a penny at a time and in manufacturing, the year-end result on the balance sheet is made on the minute-by-minute decisions that are taken throughout the year on the shop floor. Now, through the plummeting costs of small computer systems, management in the smaller companies are able to get a firm grip on those 'passing minutes' by the use of computer control systems (Greene, 1978; Williams, 1979).

When the management of a small company decide to install a computer system they often do not have the necessary in-house expertise to make the decision as to which, if any, system is best for them. They are often overwhelmed with trade literature, uncertain as to which system offers the best growth potential and in doubt as to whether to buy from one of the 'giants' of data processing, or to go to one of the specialist companies, or even whether to contract their manufacturing control problem out of their plant to a consultant systems analysis company, who will analyse their particular problem and implement a manufacturing data system to solve it.

Often, the consultants do the data processing at their own facility and the clients get the results without the expense of installing their own system, and without the need to be concerned with regular upgrading of the system to the latest state of the art.

Because of the different requirements of the different companies that will buy a given computer system, it is important that the computer system accommodates the user company, rather than the company having to accommodate the computer. This is not to say that the company should not look at its internal systems, to see if they cannot be made more efficient by the use of a methodology that makes the operator – computer interface a simple, rather than a complex, translation procedure.

One system which seems to have taken this route is the modular applications system (MAS), which has been used in over 650 facilities throughout the world (Weimer, 1977). There are three objectives behind the design philosophy of the MAS: the first is that the system should be easy to understand and use, since the world does not consist of geniuses. MAS also does not require that management need change the way in which they run their business, as is required by most MRP systems.

The second objective is for a flexible system, since as the company becomes more sophisticated, then the computer system should reflect the way in which the company wishes to work. Hence the system must develop with the company, by switching on facilities that are built into it, rather than the company having to change their computing system and having to go through the time-consuming problems of selection and debugging.

The third, and perhaps most important, objective is that the system should fit the individual company's needs. With MAS, the users get exactly the system they require, through the modular construction of the system, which is built up of elements that are built up from jobs, that come from programs that themselves come from modules.

EXAMPLES OF MICROELECTRONICS IN PRODUCTION CONTROL

In the past, control of production operations has been accompanied by a large volume of indigestible paperwork, usually late in presentation and largely inaccurate. Many computer installations served to perpetuate this state of affairs and everybody is familiar with the 5-inch thick fanfold of print-out that nobody has 3 hours of spare time to read (Hammond and Kochhar, 1978). Recent developments, particularly minicomputers or microcomputers, operating in the OLRT mode, are suitable for handling the dynamic problems inherent in production planning and control (PPC) systems. It is important to remember that, because of the many complex control functions taking place simultaneously in such systems, the right control action and data sampling intervals must be used for each control function. Thus, it is probable that personel will require data at differing times, every half hour, every day or every month dependent upon their position and what data they require at any particular time. In order that all users can be accommodated, the system must be truly OLRT, since any departure from this ideal could lead to a falling-off in the efficiency of control and could cause instability.

Some examples of current applications will now be described briefly.

(a) *International Telephone and Telegraph Co.*

OLRT process control is now a reality, with the installation of an automated manufacturing control system: based on sampled data at the International Telephone and Telegraph Co. (ITT) of Harlow, UK, who manufacture and assemble quartz crystal devices of various specifications and oscillation frequencies (Hollwey *et al.*, 1978). The application at ITT is for controlling a production line that consists of batches of crystal oscillators of varying quantities (few to hundreds). This line also accommodates a special delivery service that operates on a week's delivery for customers with urgent requirements.

Since the production line handles 1000 to 1200 batches per week and provides delivery times of 1 to 5 weeks, there can be at any one time an average of 4000 batches in operation; also a particular customer's order can contain several different items, which in turn are divided into batches of up to 25 crystals. Thus, to ship a complete order may require the successful completion of many batches, the process being further complicated by the complex technical nature of the product which necessitates providing an excess of crystals to allow for scrap losses. This is further complicated by a highly variable failure rate.

The detection of short batches, therefore, represents a major problem at ITT and yet is vital if shortages are to be made up prior to the due delivery date, since substantial cost penalties are incurred by the special delivery service, if the deliveries are late. Thus there is a need for accurate, up-to-date information flows relating to the state of every order, so that good production control is achieved.

(b) *Rubery Owen Co.*

Process control, through the application of OLRT computer systems, exist at Rubery Owen, Darlaston, UK, where the analysis of various parameters ensure consistency of spot weld quality of the company's motor vehicle wheel production, without having to resort to destructive testing (Ward, 1979).

(c) *British Sugar Corporation*

The British Sugar Corporation (BSC) operates one of the UK's most unusual and testing industrial processes: their 120-day 'campaign' during which BSC's plants will operate non-stop to process some 8 million tonnes of sugar beet. After the campaign the 17 processing plants return to the 8-month 'slumber' during which time all the machinery will be stripped down and rebuilt (Watson, 1979). For several years now the success of the campaign has been dependent upon microelectronic control systems.

The idea of using microelectronics was conceived some 5 years ago when BSC had two objectives and one big worry to resolve. In order to remain competitive, BSC wanted to obtain more control over its continuous process plants and thus increase efficiency. However, switching to microelectronics would require an investment of £150 million during the 5 years ending 1980 and, in order to justify this expenditure, BSC required that output be increased.

The big problem was how to overcome their employees' natural fears that computer technology presented a threat to their jobs, since a major strike during the 120-day campaign would be disastrous. Therefore, BSC's approach was to present MPUs as what they are, simply a new and very powerful tool to help the process operator in the complex decisions that he has to make. A retraining programme was carefully tailored in such a way that any fears of computers because of potential job losses were negated.

Having previously tried massive centralised computer systems and seen the effects of a breakdown on the process, BSC saw the advantages of microelectronics in that they allow the plant control system to be built up in modules and, in consequence, permit a simple plug-in approach to servicing.

With reference to the objectives, since the introduction of MPUs, company profits have risen steadily from £7.9 million in 1974–75 to £25.6 million in 1977–78. Turnover has increased from £115.5 million to £304.2 million, whilst at the same time, the number of employees has remained between 5500 and 6000. Output has increased from 568,248 to almost 1 million tonnes and is expected to reach 1.25 million tonnes in 1980.

Regarding the introduction of micro-electronics, BSC claim to have discovered four 'golden rules':

(1) the introduction of any technology has got to be accepted by the staff from the start;
(2) a lot of money and time has to be spent on explaining the technology and retraining staff, before the technology is installed;

(3) the system has to have the right manual back-up; in other words, traditional skills must be available to take control if the MPUs fail;
(4) the technology must not be beyond the capability of the maintenance staff.

According to BSC's experience, if all these things are right, then employees and employers alike have nothing to fear and a lot to gain from microelectronics.

(d) *East Worcestershire Waterworks Company*
The East Worcestershire Waterworks Company supplies water to a population of nearly a quarter of a million occupying an area of 300 square miles in the West Midlands of the UK, and as a result of the creation of two New Towns, together with natural expansion, the population will have increased to 300,000 by 1985 (Burkitt, April 1978).

The manpower shortage in the 1950s–1960s, arising as a result of their proximity to the industrial Midlands, forced the company to review the method of plant operation and this led to the decision to implement automatic control of all stations as an interim measure, pending the installation of more sophisticated computer-based telecontrol equipment. Automatic control was completed by 1966, thus enabling the three-shift manning to be reduced to a single shift, with the resultant reduction in manpower of three men per station. This was followed in 1970 by the telecontrol system, the justification for this system being mainly centred on a reduction in energy costs.

After 3 years of satisfactory operation, it was evident that the benefits that could be obtained from the employment of new techniques would be greater than had been anticipated. Therefore, a research project was initiated with Cambridge University for the Investigation and implementation of on-line methods of pumping plant operations. The project was started in 1974 with the prime objective of studying the feasibility of a computer control scheme which would ensure minimum cost pumping. Such a scheme was proven off line and work was begun to implement it on the network (1976).

The computer program involves the prediction of normal consumption 24 hours ahead, which is then used in an optimal control program to establish the pump schedules. The effect of on-line control upon energy costs will be assessed during the next phase of the study period and compared with the saving achieved by manual control, in order that a cost–benefit analysis can be introduced.

In terms of the previously mentioned labour problems, two interesting patterns emerged from the new operational structure:

(1) Whilst a relatively large number of semi-skilled men were eliminated from the work-force, the small team of highly skilled technicians of the calibre required to maintain the system was difficult to obtain and a company training scheme was necessary to safeguard the future position.
(2) A tremendous interest and enthusiasm for the system was shown by all concerned, and whilst the initial aim was to provide an efficient and reliable method of plant operation at minimum cost, the particular system has fulfilled an equally important requirement–that of providing a high degree of job satisfaction. This undoubtedly derives to a certain extent from the requirement–which has been met–for job enlargement.

(e) *Production control in the plastics industry*
The design of a computerised production planning and control system for the plastics products manufacturing industry requires special consideration because of a number of inherent process factors (Hammond and Kiochhar, 1978). These factors are that:

(1) The injection process lies somewhere between flow and batch production.
(2) Manufacturing routes are uncomplicated, with run times averaging 2 weeks, and only a limited requirement for complementary parts, shop loading problems are not complex.
(3) A tool tends to repeat successfully when allocated to the particular machine on which the run norms were established for the product. There can be problems, when there is a forced transfer to another similar machine, and whilst because of queueing problems it makes good sense to transfer a product to an oversized machine, the additional costs need to be quantified.
(4) The development of a new tool tends to be time-consuming and could inhibit machine availability; hence this can be an important cost factor.
(5) New business is very competitive with low yield from quotations. Dependent on policy, customers may feel trapped by tool cost transfer penalties, but within this constraint, sound products can develop a good relationship.
(6) Raw materials are a most significant element of product costs and

could underline the need to give priority to a system based on material planning, control and costing.

(7) Plant is generally run on a shift basis, but there is plenty of scope to lift machine utilisation levels, and a planning and scheduling system to make better use of resources would be the heart of any new plan.

The increasing flexibility and power of low-cost computers means that the usage of sophisticated forecasting models can now be justified. This is of great importance to the plastics products industry, since whilst a company may have an order book with firm commitments extending several months ahead, the lead times of raw material inputs may extend beyond the order book time-frame. This means that the company has to estimate its future production programme.

(f) *Factory information systems*

The UK computer company International Computers Limited (ICL) have recently launched a new range of factory information systems (Large, 1979b). The key thing about this new system is that it is a shop-floor terminal that is tough and small and has numbered touch-sensitive spots rather than tabs. On this terminal the operator and/or foreman 'tell' the central computer when a job starts and finishes or what is holding it up. The procedure for doing this is for the operative to slot a card into the terminal to announce his presence, and the terminal prompts him with questions about the job at hand.

Thus from up to 160 points in the factory or on remote sites, the system provides instant information about work in progress, the supply of parts, costing and quality control, and who turned up for work and when.

THE COMPUTER-AUTOMATED FACTORY (CAF)

As stated earlier, an integrated manufacturing system is a series of interrelated areas of responsibility operating in concert, such that synergistic benefits are derived. The CAF, which is forecast to be a reality by the end of the century, will be an embodiment of such a system. Precursors of the CAF, in the form of flexible manufacturing systems (FMSs) do, however, exist within today's manufacturing industries. The most advanced is the East German 'Prisma-2' and the Japanese 'Methodology for Unmanned Manufacture' (MUM) projects (Arndt 1977(a), (b); Hollingum, 1979b).

The Prisma-2 system is capable of automatically machining up to 72 different prismatic parts up to a size of 1.0 × 0.63 × 1.6 m, and weighing 630 kg. The system is run by six operators on the shop floor and another on the computer.

The MUM project is reportedly the ultimate FMS presently planned. A prototype plant is to be built in the early 1980s, in which some 2000 different components are to be made in batches of 1 to 25 in a factory staffed by 10 people. Processing will not only include automatic machining and assembly, but also forging, heat treatment, welding, presswork, inspection and painting.

The UK is investigating the feasibility of the unmanned batch mode production factory with its 'automated small-batch production' (ASP) project that is funded by the Department of Industry (Hollingum, 1979c).

Another indicator of the importance that has been placed on the development of the integrated manufacturing system is the USAF 'integrated computer aided manufacturing' (ICAM) project, which is to develop a completely integrated system, covering all of the computer aids to manufacturing in the aerospace industry. This project which is funded to the extent of £200 million over the 5 years ending 1982, will affect not only the aerospace industry in the USA and Europe, but wherever computer aids are used in batch manufacture (Appleton, 1977; Hollingum, 1977, 1978, 1979; Page, 1979).

Whilst the ICAM project is being conducted worldwide, so far only one contract has been awarded in the UK. This contract, initially for 1 year and worth £100,000, is under the management of the Salford University Industrial Centre and involves two project teams, one from the University of Manchester Institute of Science and Technology and the other from the University of Salford. If the UK investigators can achieve the results within the time-scale set by the USAF, a follow-on contract is expected for a further 4 years and worth an additional £1 million. The aim of the British project is to develop a computer-assisted job shop control system for sheet metal processing.

Group technology techniques are being applied to the sheetmetal working industry, and it was quickly recognised that a matrix of parts could be built into one large sheet and sheared into discrete parts later. Computer numerical controlled (CNC) punch presses that cost £110,000 are being installed at a steady rate, not the least reason being that they can do the work of five template-following turret machines. Whereas traditional sheet-metal designs economised on bends, holes and profiles, the new technology requires no such economy, yet by designing around the CNC machine's capability, direct labour costs can be £2.50 *v.* £13.00 for the same part produced by traditional methods (Page, 1979).

Where will it lead? Ultimately it will allow production only barely within our present comprehension, both managerially and technically. It is thought to allow engineers, not only to make an optimal part design, but at the same time to subject it to a performance evaluation and plan its most economic manufacture, within the confines of schedule, material availability and variability of processes and materials. The factory would be able to make a production test immediately, and automatically introduce the part into production within an established manufacturing plan.

This concept of integrated manufacturing systems, that the USAF expect to be operational by the 1990s, gives credence to the predictions of a series of Delphic forecasts made in the USA in the early 1970s (Arndt, 1976, 1977 (a), (b); Heginbotham, 1979; Levy, 1979).

In 1971 the International Institution for Production Engineering Research (CIRP), which is composed of research experts representing almost all of the production engineering research laboratories in the world, looked into the possibility of the CAF existing before the end of the century.

At about the same time two other groups of interested experts conducted their own Delphic sessions. These two groups were:

(1) a panel of US manufacturing personnel, in conjunction with the Industrial Development Division of the University of Michigan;
(2) the Society of Manufacturing Engineers in co-operation with the University of Michigan and the National Machine Tool Builders Association, and using a panel of managers and engineers from the US metal-working companies.

Whilst the three forecasts differ slightly in timing, they all forecast a strong rapid trend towards the wide development of computer-automated machine tool systems and the realisation of the CAF before the end of the century. The forecast schedule of events is:

(1) by 1980–85 computer software systems for full automation and optimisation of all steps in the manufacturing of a component will be developed and in wide usage;
(2) by 1985–88 full on-line automation and optimisation of complete manufacturing plants, controlled by a central computer, will be a reality;

(3) by 1987–90 more than 50 per cent of all machine tools produced will not have 'stand-alone' use, but will be a part of a versatile manufacturing system with automatic parts-handling between stations and controlled from a central process computer.

In the long run it is 'corporate necessity' that will bring about the CAF. This is because of the obvious economic advantages of reduced throughput time, and costs, increased machine utilisation and hence increased productivity (Arndt, 1977 (a), (b)).

General Motors (USA) are finding it necessary to adopt computer control of machines in their manufacturing and assembly plants, because they have determined that the ratio of productivity to costs has declined for the use of traditional machines (Decker, 1975).

The need to conserve energy in all of its forms is perhaps the major incentive for industries, or nations, to promote and adopt the CAF. That they will be the most efficient converter of raw materials to products can be assessed from the following:

(1) minimum wastage of expendable materials such as welding electrodes and paint;
(2) minimum scrap because, once set up, the machines never vary or make mistakes;
(3) the closed environment of the CAF, and the removal of humans enables energy savings through the usage of machine-orientated climate control and because machines do not require visible light to operate, thus there is no need for brightly illuminated workplaces and factories;
(4) because the machines are computer-controlled, the rate of material removal can be such that the optimal relationship between energy expended in metal removal, tool wear and operation time, is achieved.

In this chapter, it has been shown that a manufacturing company consists of many different areas of responsibility that must work in concert, if the company is to operate as an integrated whole with the resultant efficiency of operation.

It has also been indicated that a company or industry cannot conduct its business in a void, and that it is essential that its manner of conducting business is accepted and supported by the government and unions. Therefore it is essential that government, industry and unions go hand in hand into the microelectronics age.

BIBLIOGRAPHY

Appleton, D. S. 'A strategy for manufacturing automation' *Datamation*, vol. 23, no. 10 (October 1977), p. 64.

Ardnt, G. 'Forecasting and production technology: an outlook on manufacturing systems'. Institute of Engineering, Australia, Annual Engineering Conference, Townsville, Australia (10–15 May 1976), Paper 425/1, pp. 63–67.

—— 'Developments in computer manufacturing'. *Automation and Control*, vol. 6 (1977a), pp. 47–52.

—— 'Integrated flexible manufacturing systems: towards automation in batch production'. *New Zealand Engineering*, vol. 32, no. 7 (15 July 1977b), pp. 150–5.

d'Auria, A. and Salmon, M. 'Examples of applications of the Sigma robot' 3rd Conference on Industrial Robots and 6th International Symposium on Industrial Robots. (University of Nottingham, UK, 24–26 March 1976), Paper A2.

Bestwick, P. F. and Lockyer, K. G. 'A practical approach to production scheduling'. *International Journal of Production Research*, vol. 17, no. 2 (1979), pp. 95–109.

—— —— *Mathematics of Production Management* (Pitman, forthcoming).

Burch, R. H. *et al.* 'An advanced method of water supply and distributed network control'. *L.S.E. Engineering Bulletin* vol. 13, no. 3 (March 1976), pp. 1–21.

Burkitt, A. 'Microelectronics find a macro market in the motor industry'. *The Engineer*, vol. 246, no. 6361 (23 February 1978), pp. 51–86.

—— 'Bleak outlook for microprocessors in the motor industry'. *The Engineer*, vol. 246, no. 6365 (23 March 1978), pp. 22–3.

—— 'The microprocessor revolution could be the greatest of them all'. *The Engineer*, vol. 246, no. 6370 (27 April 1978), p. 39.

Butler, P. 'Robot electro optic eyes put to test on a conveyor belt test'. *The Engineer*, vol. 246, no. 6362 (2 March 1978), p. 49.

Chao, J. T. 'Chips impact production planning'. *Datamation*, vol. 24, no. 2 (February 1978), p. 41.

Charette, W. P. 'Making project management easy'. *Datamation*, vol. 24, no. 4 (April, 1978), p. 47.

Collins, J. 'Why more attention should be given to assembly work'. *The Engineer*, vol. 248, no. 6413 (22 February 1979), pp. S61–S64.

Cox, J. L. 'Minicomputer systems upgrade factory management'. *IEEE*

Transactions on Manufacturing Technology, vol. MFT-6, no. 2 (1977), pp. 20–2.

Decker, R. W. 'Computer aided design and manufacturing at GM'. *Datamation*, vol. 24, no. 5 (May 1978), p. 159.

Dudley, N. A. 'Comparative productivity analysis – study in the United Kingdom West Midlands engineering and metalworking industries'. *International Journal of Production Research*, vol. 8, no. 4 (1970), pp. 397–403.

—— *Midlands Tomorrow*, no. 8 (1975). Issued by the West Midlands Economic Planning Council, UK.

Elliot, M. 'Fears and hopes of the MPU revolution'. *Electronics Weekly*, no. 957 (31 January 1979), p. 5.

Engelberger, J. F. 'Performance evaluation of industrial robots'. 3rd Conference on Industrial Robots and 6th International Symposium on Industrial Robots. (University of Nottingham, UK, 24–26 March 1976), Paper J4.

Forslin, J. and Söderlund, J. 'Automation and the work organisation – a case study from the automobile industry'. Personal copy of case study from Volvo, Sweden.

Greene, A. M. 'Entry-level computers attract manufacturing'. *Iron Age*, vol. 221 (1 May 1978), pp. 42, 45.

Hammond, A. and Kochhar, A. K. 'The role of computers in production planning and control of plastic products'. *Plastics and Rubber: Processing*, June 1978.

Hartley, J. 'The changing scene in car manufacture'. *Automotive Engineer*, vol. 2, no. 2 (April/May 1977), pp. 14–19.

—— 'A growing family to cut welding drudgery'. *The Engineer*, vol. 246, no. 6361 (23 February 1978), pp. 44–5, 47–8.

—— 'VW robot family'. *Automotive Engineer*, vol. 3, no. 2 (April/May 1978), p. 68.

—— 'Robots in the automotive industry'. *Automotive Engineer*, vol. 3, no. 5 (October/November 1978).

—— 'Japanese aim for more flexibility'. *The Engineer*, vol. 248, no. 6407 (11 January 1979), pp. 32–37.

—— 'Letting robots take the overtime strain' *The Engineer*, vol. 248, no. 6408 (18 January 1979), pp. iv–vii.

—— 'Now Renault offshoot is to market a see-all robot'. *The Engineer*, vol. 248, no. 6409 (25 January 1979), p. 15.

—— 'Renault takes the robpt road'. *The Engineer*, vol. 248, no. 6413 (22 February 1979), pp. S27–9, S32–3.

Haugan, K. M. 'Continuous path controlled robot in arc welding application'. 3rd Conference on Industrial Robots and 6th International Symposium on industrial robots. (University of Nottingham, UK, 24–26 March 1976), Paper G2.

Heginbotham, W. 'The future lies in more capital intensive solutions'. *The Engineer*, vol. 248, no. 6413 (22 February 1979), pp. S14–15, S17.

Heisel, J. A. *et al.* 'Microcomputer controlled resistance welder'. *IEEE Transactions on Industrial Electronics and Control Instrumentation*, vol. IECI-25, no. 1 (February 1978), pp. 14–16.

Hollingum, J. 'Integrated computer system for aerospace'. *The Engineer*, vol. 245, no. 6340 (29 September 1977), pp. 52–3, 55.

—— 'USAF cash aid will boost small British engineering firms'. *The Engineer*, vol. 247, no. 6400 (23 November 1978), pp. 42–4.

—— 'The collaborators – and it does pay off'. *The Engineer*, vol. 248, no. 6413 (22 February 1979a), pp. 36–7, 41.

—— 'Trend overseas moving towards more automation'. *The Engineer*, vol. 248, no. 6413 (22 February 1979b), pp. S37–9.

—— 'Automation will grow along with computing power'. *The Engineer*, vol. 248, no. 6413 (22 February 1979c), pp. S53, 56–7.

Hollwey, M. W. M. *et al.* 'Using a real time mini-computer for production control'. *Production Engineer*, vol. 57, no. 11 (November 1978), pp. 37–42.

Hosier, J. 'Word processing'. *Administrative Management*, vol. 28, no. 3 (November 1978), pp. 20–1.

Hyder, K. 'The new wealth of nations is in information'. *Labour Weekly*, no. 355 (10 November 1978).

Jephcott, B. 'Robots pave way for 20 new jobs'. *Birmingham Evening Mail*, (27 March 1979), p. 34.

Jones, F. 'Authorities should look to word processing'. *Electronics Weekly*, no. 975 (6 June 1979), p. 6.

—— 'Word processing – the Bradford way'. Conference script-code No. AMF0051010/JCM.

Kerr, J. 'Only the robots know what the foundries of Britain are missing'. *The Engineer*, vol. 246, no. 6366, (30 March 1978), p. 39.

Kochhar, A. K. 'Distributed real time data processing for manufacturing organisations'. *IEEE Transactions on Engineering Management*, vol. EM-24, no. 4 (November 1977), pp. 119–24.

—— 'The use of computers and analytical techniques for production planning and control in the British manufacturing industry'. *Computing and Industrial Engineering*, vol. 2, no. 4 (1978), pp. 163–79.

—— and Parnaby, J. 'The choice of computer systems for real time production control'. *Production Engineer*, vol. 56, no. 10 (October 1977), pp. 29–33.

Laerdal, S. 'Trallfra robot – A new concept of the control system'. 3rd Conference on Industrial Robots and 6th International Symposium on Industrial Robots. (University of Nottingham, UK, 24–26 March 1976), Paper G3.

Large, P. 'Chips make Britain first – a "de-developed" nation'. *Financial Guardian*, 2 May 1979a, p. 20.

—— 'More micro-mothers move in'. *Guardian*, 6 June 1979b, p. 17.

Levy, G. W. *et al.* Final report on The Manufacturing Engineer – Past, Present and Future, to Society of Manufacturing Engineers. Columbus, Ohio (Battelle Columbus Laboratories), 1979.

Lockyer, K. G. 'Production management: the unaccepted challenge'. Inaugural lecture delivered at University of Bradford, 28 October 1975.

McGann, S. (S. M.) 'Equipment for automobile production'. *Automotive Engineer*, vol. 2, no. 2 (April/May 1977), p. 20.

Marley, J. 'Evolving microprocessors which better meet the needs of the automobile electronics'. *IEEE Proceedings*, vol. 66, no. 2 (February 1978), p. 142.

Mitchell, J. 'Which word processor'. *Administrative Management*, vol. 28, no. 3 (November 1978), pp. 22–3, 27.

Mortimer, J. 'Britain way behind in robot use'. *The Engineer*, vol. 246, no. 6362 (2 March 1978), p. 17.

—— 'Time for a fresh look at shop floor thinking'. *The Engineer*, vol. 248, no. 6413 (22 February 1979), pp. S39–41.

Myers, E. (E. M.) 'Voice input – where it stands'. *Datamation*, vol. 24, no. 5 (May 1978), p. 274.

New, C. C. *Managing Manufacturing Operations* (London: BIM, 1976; British Institute of Management Survey Report No. 35).

Page, M. 'Punching sheet metalworking into better shape for everyone'. *The Engineer*, vol. 248, no. 6413 (22 February 1979), pp. S67, 70–1.

Schraft, R. D. and Schmidt, V. 'A computer aided method for selection of an industrial robot for the automation of a workplace'. 3rd Conference on Industrial Robots and 6th International Symposium on Industrial Robots. (University of Nottingham, UK, 24–26 March 1976), Paper A2.

Sheldon, R. P. Personal copy of unpublished internal report of the British Association for Advancement of Science conference 'Automation – Friend or Foe?' (Bradford, March 1979).

Sheridan, T. B. 'On human supervision of industrial robots'. 3rd Conference of Industrial Robots and 6th International Symposium on Industrial Robots. (University of Nottingham, UK, 24–26 March 1976), Paper A4.

Simmonds, W. H. '*Microprocessors and their Application*'. (Sira Institute Ltd., UK, November 1978).

Stonier, T. 'The rise and rise of the knowledge industry'. *The Business Location File*, June/July 1978, pp. 5–6.

—— 'On the future of employment'. *N. U. T. Guide to Careers Work* (1979), pp. 30–1.

Thornton, P. 'Employment implications of the new technology in Yorkshire and Humberside'. Text of paper given at 'The Impact of New Technology on the Regional Economy' conference (Leeds Polytechnic, UK, 15 May 1979).

Toffler, A. *Future Shock*. (Pan, 1971; 15th printing 1978).

Trades Union Congress (TUC). *Employment and Technology – a TUC Interim Report*. (TUC, 1979).

Tyler, C. 'The TUC nibbles at new technology'. *Financial Times*, 18 June 1979, p. 17.

Ward, D. 'Where to now, R2D2?' *Metalworking Production*, March 1979, pp. 102–3, 107.

—— 'Next stop for computer power – the shopfloor'. *Metalworking Production*, March 1979, pp. 129, 133, 137.

Watson, I. 'How BSC sugared its chips'. *The Sunday Telegraph*, 21 January 1979.

Weimer, G. A. 'Computerized controls grow in smaller companies'. *Iron Age*, vol. 220 (5 December 1977), pp. 35–7.

Williams, E. 'Study to look at more use of electronics'. *The Engineer*, vol. 248, no. 6413 (22 February 1979), p. 12.

—— Personal letter dated 11 May 1979.

Wood, L. D. 'Distributed processing in manufacture'. *Datamation*, vol. 23, no. 10 (October 1977), p. 60.

Anon. 'Computer automated factories predicted a reality in this century'. *Digital Control and Automation Systems*, vol. 16, no. 9 (1977), p. 76.

—— 'GM's PUMA robot tackles assembly line chores'. *Iron Age*, vol. 221 (5 June 1978), pp. 12–13.

—— 'Chips presence is felt at IDEA show'. *Electronics Weekly*, no. 956 (24 January 1979), p. 7.

—— 'New jobs – that's the Servis spin-off'. *Birmingham Evening Mail*, 27 March 1979, p. 34.

8 Management Systems and Communications

M. F. WOODS

INTRODUCTION

Management systems have been with working managers ever since the experimental entrepreneurship of the industrial revolution became codified. However the advent of the main-frame computer after World War I, and the success of operational research, provided new impetus to their extension. With several, and notable, exceptions the move towards computerised management systems has in practice proved slow and disappointing.

In this section of the book we will seek to explain the comparative failure of many computerised management systems in behavioural terms, and show how a new generation of systems – many of which are near relatives of the old systems, but based on the microcomputer, are likely to achieve a much wider acceptance.

The chapter begins with a survey of the manager's job and attempts to show how the new 'micro' management systems allow for much greater personalising on the part of the individual manager and decision-maker. It closes with a warning that the changes the 'micro' systems are likely to bring about may not be in the currently fashionable direction of liberalising business. They may make the manager and decision-maker much more remote and authoritarian.

THE MANAGER'S JOB

Any management system has to fit the role of management as perceived by active managers. This point although almost axiomatic, seems to have been missed by many 'devisers of systems'. In 1973 Mintzberg reported a study on the nature of managerial work. Since 1973 several workers (Kurke *et al.*, 1979; Stogdill, 1974; Tornow and Pinto, 1976; McCall *et al.*, 1978; Morrison *et al.*, 1978) have rechecked and enlarged on the study and confirmed the principle of Mintzberg's findings. In 1980 McCall and Segrist published a definitive paper which I shall use as a starting point for this chapter.

> The job is fragmented by many brief episodes. The pace is staggering and constant. Most of the time is spent talking to people from all over the organisation and its environment. The trivial and the unimportant intermingle unpredictably; form and substance shadow-box throughout the day. Variety predominates. Activities, contacts and topics range dramatically.

This vivid description from McCall, Morrison and Hannan's work on management diaries brings into focus the repeated cries of academics that management tools are not being applied sufficiently.

Mintzberg drew out ten management functions or roles:

> figurehead, leader, liaison, monitor, disseminator, spokesman, entrepreneur, disturbance handler, resource allocator, and negotiator.

He points out that while there is a cluster of activities formed round each of these functions, there is major overlap: 'Moreover, the ten roles constitute a gestalt of managerial work which loses its character if one of the roles is omitted' (McCall and Segrist, 1980).

McCall and Segrist, in a distinguished statistical study, showed that these ten factors reduced to six on analysis and that the ranking of the six in order of importance against three job categories and six levels in the organisation were surprisingly consistent (Table 8.1). Leadership and entrepreneurial functions were predominantly regarded as 'the job'. The two 'communications' factors – liaison and spokesman, came 3rd and 6th, and the 'decision' factors – monitoring and resource allocation, can be almost equal with slight variations due to grade and job. These figures are less surprising considering the mean duration of any activity in the total sample was 8–14 minutes.

The importance of these findings to our study of management systems and communication against the impact of microelectronics may not be too obvious. The manager sees him/herself as an entrepreneur and a leader – *all* other functions are subordinate to these two. Any manage-

TABLE 8.1 *Managers' appreciation of their own job functions in order of importance (after McCall and Segrist, 1980); mean scores and rank by level and decision*

	Job Area																	
	Manufacturing Seniority, High→Low						*Sales Seniority, High→Low*						*Finance Seniority, High→Low*					
Ranking of Function	*1*	*2*	*3*	*4*	*5*	*6*	*1*	*2*	*3*	*4*	*5*	*6*	*1*	*2*	*3*	*4*	*5*	*6*
Leader	1	2	2	2	2	2	1	1	2	2	2	4	3	2	2	2	2	4
Entrepreneur	2	1	1	1	1	1	2	2	1	1	1	1	1	1	1	1	1	1.5
Liaison	6	6	6	6	6	6	6	6	6	6	6	5	6	6	6	6	6	6
Monitor	5	5	5	5	4	4	5	5	5	4.5	3	2	4	4	4	5	5	5
Spokesman	3.5	3	4	4	5	5	4	3	4	4.5	4	3	2	3	3	3	3	3
Resource allocator	3.5	4	3	3	3	3	3	4	3	3	5	6	5	5	5	4	4	1.5

NOTE: Numbers *1–6* in column-headings signify job functions listed in column 1.

ment system that subverts these roles towards making the others more efficient is likely to fail. There is an image that fixes in the mind when we use the concepts of manager as leader and entrepreneur – the image is that of isolation and privacy. We must ask ourselves how the surge of techniques arising from academics in the 1960s and 1970s met this image; I would maintain – not very well.

Let us have a look at the resource allocation–network project systems, for example. On the face of it they gave managers precise data – based on their own estimates, on the function managers see as their third most important job function – resource allocation. How was the system applied in practice in the 1960s and 1970s?

(1) The manager was given the basic information that the system exists and 'can be moulded' to his /her requirements by a third party – consultant internal or external, a business school, or perhaps a book.
(2) After a relatively laborious process the 'system' is adapted to the manager's specific needs and put on a main-frame computer somewhere. The process involves a quite challenging analysis of his/her particular needs – or as it is perceived – management style.
(3) The system runs, through an intermediary, and supplies the data developed as expected from the first analysis.

The whole of this process is public and involves intermediaries. The manager is *very* unlikely to have either the time or the ability/skills to understand how the 'thing' works. It is very unlikely he will be able to communicate effectively with the 'main-frame' specialists on whom he has suddenly become totally dependent. Minor changes in his/her requirements are likely to be ignored because of the effort required to communicate them. The main-frame experts are seen, at least on an emotional level, as a threat and everything they generate as a challenge to the two principal points in the hierarchy by which he judges his/her work – as an entrepreneur and a leader. The 'system' has attempted to usurp the role and is resented and ultimately rejected.

An alternative scenario we know of was of a highly numerate senior manager who attempted to pass on a highly personalised 'system' to his successor when he retired. The 'system' and its acolytes were not accepted as an inheritance.

Our burden of argument is that the micro-based computers and information systems now flooding into business and finding resting places on executive desks are *not* a threat to the entrepreneur and leadership perception of managers. They are under his or her control. They work in privacy. They can be taken home away from interruptions. They can accept changes of mind without raised eyebrows. They are a servant and never a critic. In the jargon of our age:

They are user-friendly.

By allowing privacy in decision-making, they allow truth, for truth can only exist in privacy.

THE ENVIRONMENT OF THE MANAGER

The second starting point of argument for the introduction of microelectronics into management systems and communication is concerned with change. Again we will begin by using one single and much quoted reference of the past – the work of Emery and Trist (1960) on socio-technical systems. The two authors argued 20 years ago against the use of closed systems thinking. The present author found then and now their historical approach valuable, and will use it here.

> We have now isolated four 'ideal types' of causal texture, approximations to which may be thought of existing simultaneously in the 'real world' of most organisations – though, of course, their weightings will vary enormously from case to case.
>
> The first three of these types have already, and repeatedly, been described. . . . The fourth type, however, is new, at least to us, and is one that for some time we have been endeavouring to identify (Emery and Trist, 1965).

The four types of 'causal texture' are:

(1) *The placid randomised environment.* A manager in such an environment would be asked to 'do one's best on a purely local basis'. A vision of a caveman living on a vast plain, or a single entrepreneur in a new country, comes to mind.
(2) *A placid clustered environment.* Survival depends on strategy and part of the strategy must be the clustering of resources. A feudal system society is the historical analogy and the large hierarchial organisations of the 'easy years' of the industrial revolution are relevant.

(3) *A disturbed–reactive environment.* Here interactions between organisations become important and in history we come into the areas of the military theorists. In organisations we have the golden age of operational research and closed system thinking. We accept that 'what goes in goes out' and the problem is optimisation of resources – human and financial. Modelling methods work and strategies based on co-operation or cornering can be developed. Moving back to the subject of the chapter – here, in the 'disturbed–reactive environment', the main-frame computer systems had their full application.

(4) *Turbulent field environments* – unlike (3) systems we do not merely have to optimise 'what goes in goes out'. There is an interaction with the environment itself. As Emery and Trist put it, 'The ground is in motion'.The concept of turbulence in organisations was new in 1965 and even their example in the article of a food-canning company whose environment moves in a highly complex way, seems, slightly forced.

To a manager of the 1980s the concept of turbulence must seem very familiar. 'Fine we've got all the order – all the raw materials and the process works. Now they've called a national day of protest so we have to reschedule from scratch.' This is the world we now live in.

Let us summarise:

(1) the management systems of the 1960s and the 1970s were strangely public from the viewpoint of active managers;
(2) the very perception of the management role by individual managers is threatened by this lack of privacy;
(3) the world of today is turbulent – reactive. The manager needs all the help he can get for survival.

The micro as personal computers or as personal information systems are tailored for the new world.

THE MICRO NOW IN MANAGEMENT SYSTEMS AND COMMUNICATION

Looking at the current use of micros in systems and communication we have no shortage of material. It is an exploding field, rather like radio in the 1930s or television in the 1950s. Anything seems possible and the

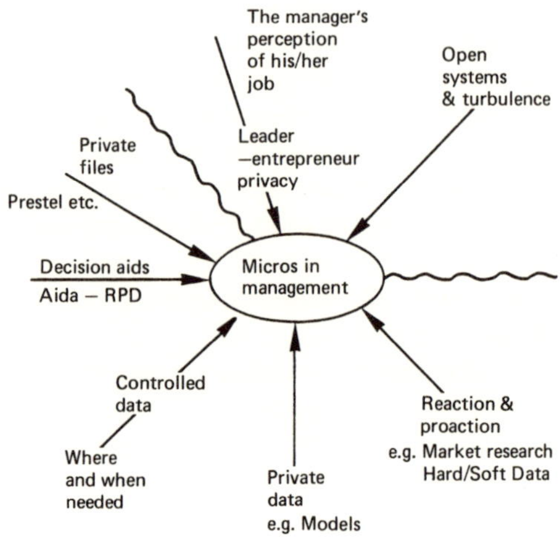

FIGURE 8.1 Summary of Micros in Management Systems and Communication 1980

exciting people who seem to be hidden underneath the carpet of less dynamic times are setting up new businesses to exploit the technology. In Figure 8.1 I have attempted to condense the present picture against the driving forces we have already discussed.

REACTION AND PROACTION

Example – market research
Market research data-collection is a typical public system on the lines we have discussed.

> 'How much of Brand X will be sold tomorrow, or how much was sold last week?'

The information is vital both to manufacturers and retailers, but it is either soft – an informed guess, or historic. Amazing techniques have been used to find better data including the purchase of demographically selected dust bins to sort wrappers. Now we can answer a much better question:

> 'How much of Brand X was sold in High Street Y TODAY?'

A wand connected to a checkout computer can give the answer where it is needed and when it is needed *factually*. There is no further need for theory – we can test cheaply how campaigns affect sales and how changes are perceived by the only behaviour that counts – point of sale SALES. The speculation that springs to mind is, of course, the value of the data and its accessibility to the retailer (as of right) and the manufacturer (by permission). Looking at this example, it obviously permits proactive management and is relevant to a turbulent society. Individual managers have the privacy of their personal data – without passing through the notebooks of market research organisations.

We can also see individual managers using the system to test theories and using shops, areas and regions as models.

PRIVATE DATA – MODELS

Example – uncertainty in bidding for contracts
Simulations or mathematical models of systems have been around for a long time. The problems have always been:

(1) the 'weight' of the hardware. Main-frame models are usually not for the uninitiated and are subject to the personalisation of working in a cathedral;
(2) the basis of the model is surrounded in complex computer language and has values unknown to the user.

Many models have now been developed for personal computers. The example we will quote is by Brian Fine (1976) and simulates the uncertainty factors in bidding for contracts. Using the model on a 32K Pet (Commodore) desk computer costing in total £1000, the individual manager can discover, in secure privacy, the variables of his proposed strategies. Prized company dictates such as: 'never build more than 30 houses on one contract' can be tested. If they are wrong nobody need know – the changes can be gradual and checked silently.

Because the microcomputer fits happily in the boot of a car, the work can be done at home. 'Wrong' results are no threat because the 'results' have the impermanence of an electron beam on a phosphor.

CONTROLLED DATA — WHERE AND WHEN IT IS NEEDED

Pope Urban is said to have explained his persecution of Galileo not on the grounds of disliking hearing the 'facts' Galileo was producing, but that everyone else *knew* he was hearing the facts.

Managers need the truth, but because they get their data publicly, they rarely get it. One major function of leadership, as no doubt Urban knew, is to be in control of the truth. To be able to adjust, and adjust-to, new data is essential to maintain many leadership situations. You may know that the factory is in a bad way. *They* may think the factory is in a bad way, but if they know you know – then there's trouble.

Truth can only exist in privacy.

Going back to our example of the supermarket checkout analysis. The manager can hear or not hear. He can store, file, collate or listen *now*. He, and he alone, is in charge. The leader and the entrepreneur are safe.

DECISION AIDS

AIDA and RPD

In the golden age of operation research many main-frame decision aids were developed. These devices were able to act as templates for management decisions. Notable was AIDA – Analysis of Interrelated Decision Analysis, by Morgan of the Tavistock Institute (Morgan, 1972; Woods and Davies, 1971). Basically AIDA took the facts of complex decision problems and reduced them to manageable proportions. The disadvantages of the system were of course those we have already discussed.

The micro version is slower and smaller but absolutely accessible. The individual can procrastinate and vacillate on his own desk.

Whether AIDA – or MAUDE in its micro version (Woods, in press) will be used is open to question but it and its fellows may change management decision-making for the good.

The Research Planning Diagram (Woods and Tweedie, 1976) similarly is a decision tool allowing managers to look at complex decision-making in uncertainty, and will soon be available on micro. Real decisions can be tested in privacy on the desk or in the home. The information, or lack of information, the risk, the alternatives and the contingencies need not be shared.

PRIVATE DATA FILES

Example – Prestel (see Figure 8.2)

Reading the words of the Prestel user's guide and seeing the 'welcome to Prestel' come onto our screens we felt the full impact of all we have said to date. The copy of the 'How to Use Prestel' is midway between the

Calling Prestel

Switch on your television set

Make sure nobody is using the telephone

Plug in your key pad (if yours is the plug-in type)

Switch to Prestel according to the operating instructions supplied with your set. After a few seconds the Prestel page welcoming you to Prestel should appear on your screen

If the page does not appear, please try again

If you still have difficulty, refer to the 'Problem?' section page(s) further on

FIGURE 8.2 A copy of the screen when calling up Prestel

spy's instructions for setting up a radio in an attic and a reassurance that you are the leader and entrepreneur of your dreams. *You are in charge.* What you command is equally remarkable. Instant data without outside help – librarians, technical journal editors, specialist departments, need to worry. Obviously Ceefax, Oriele and the French systems give the same feeling of power, but this is what the microelectronic revolution is about in management systems and communication – *power* for the individual manager.

CONCLUSIONS

The microcomputerised management systems differ in kind from their predecessors in that they are private. The information they give, the help they provide, and the structuring they perform is personalised and personal. The struggles that are inevitable in the development of any policy-making tool can happen in a time-scale unknown outside the manager's own mind. Blind alleys vanish for ever. We can envisage a microprocessor travelling with our senior manager of the mid-1980s as perfect assistant: always available, always faithful, always exclusive. If we consider the fashionable trends in management style of the 1970s as being towards participation and consultation – the open style of management, we will see a conflict. The micros, both literally and figuratively, are the confidante of one man – or woman. One can almost hear the words at a Board Meeting:

> Decision A is the best choice – it would be a little difficult to go through all the calculations here and now but if you check the figures you will see that I am right.

The structure of business that will evolve towards this style of management will certainly not be consultative.

A second, and perhaps more revolutionary, thought is that the micro has opened the door to the long-awaited divorce of the manager from a specific place of work. The micros can work anywhere and transmit anywhere. As well as the social structure of business the micros will affect the architectural structures of our businesses. The massive centralised office block is based on the construct that access to people is essential. By the late 1980s it is perfectly possible that this construct will be invalid.

REFERENCES.

Emery, F. E. and Trist, E. L. *Socio-Technical Systems.* Management Science Models and Techniques, vol. 2, pp. 83–97. (Oxford: Pergamon Press).

Emery, F. E. and Trist, E. L. '*Causal textures of organisational Environments*'. *Human Relations*, vol. 18 (1965), pp. 21–32.

Fine, B., in Turin D. A. (ed.) *Economics of Construction Management*, 1976 Godwin pp. 202–21. (Fine and Curtis, Consultants in Planning, Computing and Management Science, 18 Cotswold Gardens, London NW2).

Kurke, L. B. and Aldrich, H. E. 'Mintzberg was right: A replication and extension of *The Nature of Managerial Work*'. In R. Osborn (Chairman), '*Managerial Roles in Formal Organisations*'. (Symposium presented at the meeting of the Academy of Management, Atlanta, August 1979).

McCall, M. W. Jr., Morrison, A. M. and Hannan, R. L. *Studies of Managerial Work: Results and Methods* (Tech. Rep. No. 9) (Greensboro, N. C.: Center for Creative Leadership, May 1978).

McCall, M. W. and Segrist, C. A. *In Pursuit of the Manager's Job* (Tech. Rep. No. 14) (Greensboro, N. C.: Center for Creative Leadership, March 1980).

Mintzberg, H. *The Nature of Managerial Work* (New York: Harper & Row, 1973).

Morgan, J. R. *AIDA: A technique for the Management of Design* (London: The Institute for Operational Research, 1972).

Morrison, A. M., McCall, M. W. Jr. and DeVries, D. L. *Feedback to Managers: A Comprehensive Review of Twenty-four Instruments* (Tech. Rep. No. 8) (Greensboro, N. C.: Centre for Creative Leadership, March 1978).

Stogdill, R. M. *Handbook of Leadership.* (New York: Free Press, 1974).

Tornow, W. W. and Pinto, P. R. 'The development of a managerial job taxonomy: A system for describing, classifying, and evaluating executive positions.' *Journal of Applied Psychology*, 61 (1976), 410–18.

Woods, M. F. and Davies, G. B. 'The use of Attribute Listing in process scale-up decision-making.' I. Chem. E. Symp. Series No. 35 (London, 1971).
Woods, M. F. 'MAUDE', *Journal of Planned Innovation* (in press).
Woods, M. F. and Tweedie, T. P. S. *R & D. Management*, vol. 7, no. 1 (October 1976). Tactical Control of Research Projects using Research Planning Diagrams.

9 The Word Processor – a case study in introducing a microelectronic system

FRANK JONES

INTRODUCTION

Most organisations will, between now and the mid 1980s and early 1990s, change their modus operandum to take advantage of new techniques using microprocessor-based technology.

Some companies will see this as a challenge to grasp new opportunities, others as a necessity to remain competitive and solvent; whatever the reason the change will be apparent rather more quickly than most organisations would normally like. New management styles, new jobs, new training methods, could mean the instant redevelopment of people to the challenges required to optimise the use of microprocessor equipment. However, people are predictably reluctant to change.

> The mind is slow in unlearning what it has been long in learning (Seneca)

Between 1977 and 1978, following four man-years research, my team and I installed, in the City of Bradford Metropolitan Council, a large screen-based shared-logic word processing system, purchased from AES Wordplex. Technical details of this system can be found in Appendix A. The facts gathered from, and my experience of, this installation form the basis of the issues I am going to discuss in this chapter on the application of word processing and its impact on people.

IMPLEMENTATION

The implementation of office automation equipment will vary dependent on a number of unique conditions that exist in organisations at the time of the change. These conditions, which I shall describe as the 'MALLET', will be the determining factors in the method of installation and development of the system. These may be summarised as follows:

M Management's attitude to change;
A Attitude of staff to change;
L Length of negotiation;
L Learning period required by staff;
E Experience;
T Trade union attitude to change.

MANAGEMENT'S ATTITUDE TO CHANGE

The first hurdle for any innovator within an organisation is to get the power he requires to carry out the changes which he envisages will be beneficial to the organisation. The acquisition of the appropriate power usually requires convincing and educating his immediate superior. This was difficult, for example, in the pioneering days of screen-based shared-logic systems for word processing, as no comparable sites were available for inspection for proof of the success of adopting certain procedural, behavioural and equipment changes. Therefore, the convincing of management may need to be achieved by other means such as a presentation of ideas or training seminars.

It has been my experience that the hoarding of power is a trait most superiors have, particularly the less innovative and unimaginative

> Next to the assumption of power was the responsibility of relinquishing it (Benjamin Disraeli).

It is easier for a manager to delay projects to a pace where he is able to absorb information, than it is to give a subordinate his head and allow him to appear cleverer to their collective peers. This is why many office automation projects take a long time in coming to fruition; not trade union militancy, but management insecurity. In circumstances of this nature, the subordinate must allow the superior manager a share in the potential glory of the grand design; perhaps even allow him to appear to be the instigator. The subordinate may well feel frustrated by this suggestion, but, unfortunately, it is usually the only method by which success may be achieved in gaining the support of some managers. Some typical ploys are to draft articles for the superior to re-write and publish as joint ventures, to write lectures for the superior, etc.

If participation cannot be achieved by the superior then he should be bypassed by reference to his controlling body. This may be a Board of Directors, Panel of Councillors or a single person. The simplest method of achieving this type of opening is to organise a visit to a site where similar equipment has been installed. If the superior's attitude changes at the site visit, seminar or meeting, the subordinate should allow him to extradite himself by allowing him to take a leading role.

If, however, the superior is of the defensive kind, and is used to innovators who occasionally rock his bureaucratic domain, he may attend all seminars alone and read all submissions to his peers. A number of devices may be used to unhinge him; flood him with urgent work each day for a month until he requests you carry out non-urgent work (which will naturally be your seminar/paper/meeting with his boss); include indescribably technical papers he is unable to understand (so that you accompany him for reference) build into papers some errors that your peers are likely to spot but he may not (so that you will be called to account for your mistake at the meeting).

After all these methods have failed to lead your superior to the light, my advice is either to find another, more progressive, employer or suddenly become patient.

> We may convince others by our arguments; but we can only persuade them by their own (Joubert's 'Penser', No. 106).

ATTITUDE OF STAFF TO CHANGE

When change is first mooted staff will entrench themselves in experience and adopt attitudes which reflect how they, or somebody they know, was affected by change. They may be very receptive, arguing that their lot is a poor one now and can only improve.

> To some will come a time when change itself is beauty, if not heaven (E. A. Robinson, *Llewellyn and the Tree*).

They may remain neutral and develop an attitude as the change affects them. More than likely, however, they will be defensive and seek reassurance that the trauma of change will at least be offset by an improvement in some aspect of their work (e.g. service, pay, conditions, status, education). The most productive method of convincing staff who are prepared to allow arguments and facts to affect their attitudes, is to provide evidence of success elsewhere.

In-house seminars, at which outside speakers from successful sites participate, and the showing of films, slide and tape presentations and the publication of case studies, is one method. Familiarity with the equipment erodes controversy, and allowing equipment sampling on site by renting equipment is a useful exercise.

> When a thing ceases to be a subject of controversy, it ceases to be a subject of interest (William Hazlitt).

However, when dealing with an established decision, do not allow new interest and enthusiasm of staff to re-open the debate, or question the suitability of the decision taken at an earlier date.

> Debate destroys dispatch (Sir John Denham).

LENGTH OF NEGOTIATIONS

If negotiations with the trade union (or in the convincing of management) are protracted, the earlier work of familiarising staff with the equipment will be eroded. This is particularly the case in areas of high labour turnover such as clerks, typing pools, cashiers, etc. With the passage of time, audience diversity will be difficult to manage. Established staff who have been concerned with the project will very quickly lose interest or become suspicious if repeatedly told the same or similar information over a long period. Alternatively, new staff will not be able to understand the reasons for agreed decisions between management and staff if the background is not explained to them.

Experience in this area has allowed me to develop a simple approach to status negotiations. Be open and frank, if undiplomatic, with staff, management and unions (but of course speak unofficially). A good three-point plan to adopt is:

(1) have a status telephone extension number anybody can ring for the latest developments in negotiations;
(2) give background information separately to new staff by means of interview, slide and tape presentation or induction course (ensure

this is not a one-sided viewpoint and involves other interested parties);
(3) have meetings with all staff only when you have something new to say.

LEARNING PERIOD REQUIRED BY STAFF

The learning period may be perceived to be anything from a few days to several months. Ensure that what you perceive as an average worker is revered to the staff as an exceptional or unusual worker. Staff who lose confidence in the early stages also lose enthusiasm for the project and become a disruptive influence on other staff, so build confidence by success.

EXPERIENCE

Experience from previous occupations, qualifications, etc., is not gained easily by most staff. In the early days of the project, build reassurance into their future tasks by introducing or highlighting the relevance of existing skills; e.g. word processing:

> The skill of a screen-based word processing operator is partly dependent on her existing QWERTY keying skill and command of English.

TRADE UNION ATTITUDE TO CHANGE

Should a particular trade union already have a documented policy for implementing office automation, such as the APEX algorithm on page 21 of their report on Word Processing (reproduced in Appendix B), then a basis exists for commencement of negotiations. However, most trade unions have either left it to local branches or not yet developed national policies. The best means of progress in these circumstances is to ask the trade union to draw up a timetable for implementation and a list of objections against immediate installation. Alternatively, management could provide a similar paper for discussion, or you could draw up independent papers for comparison. You now have a basis for negotiation and a timescale to contract or extend.

> A modesty in delivering our sentiments leaves us a liberty of changing them without blushing (Thomas Wilson).

MALLET SUMMARY

The conditions in any of the MALLET areas may pre-ordain a particular development programme. For example, at the City of Bradford Metropolitan Council, the following conditions existed in 1976:

- M upper management highly sceptical of success but publicly committed to project;
- A majority unsure of benefits of change and separated by attitude and geography;
- L protracted negotiations had meant many new staff were ill-informed and trade unions and personnel office prevented free flow of information;
- L anticipated learning curve seemed to be about 12 months;
- E existing QWERTY keying skills and command of English were relevant to immediate success;
- T the trade union did not dismiss progress; only uncertain of its success.

I elected, therefore, to develop a system which collected them geographically together, was highly reliant on existing skills and would be immediately successful. The system was a 'partial content approach to word processing' and called the 'U turn shared-logic system'. This proved successful not merely because the procedures and equipment were appropriate but because the approach reflected the conditions suitable at the time of implementation.

SOME SIMPLE RULES FOR CHANGE

Having decided what the changes to the organisation are to be you need to formalise the change. Some simple rules for the change are summarised below:

(1) have an objective against which success can be measured;
(2) have a plan (Appendix C contains a typical implementation plan);
(3) consult and communicate the plan;
(4) implement the plan;
(5) monitor the success of the plan.

CHANGES MADE BY OFFICE AUTOMATION

The first achievement in successfully implementing your office automation project is in identifying the changes made. An example of the changes made by a partial content approach to word processing using a particular shared logic system may be seen below:

AUTHORS	composition reduced tasks made systematic content reduced input formalised output formalised
TYPISTS	shorthand reduced organisational changes made retraining required contact with authors reduced
OUTPUT	more rigid layout more consistent printing changes to typeface continuous stationery used.

DECISIONS AND CONSULTATION

It is important here to establish the difference between 'areas of decisions for communication' pre-ordained by earlier conclusions, and 'areas of consultation'. Naturally, different contributors to any working group may view them differently, but it is fundamental to the success of projects that there exists a core of decisions for communication to staff. The anarchy of debate where no decisions exist will lead to interminable conflicting consultations. An example of areas of decision and consultation for the partial content word processing installation may be seen in Figures 9.1–9.3.

SOME RULES FOR CONSULTATION

(1) Discuss areas of consultation, *not* decisions for communication.
(2) Avoid groups of mixed interests
(3) Make decisions with individuals, *not* working parties.
(4) Make decisions at meetings, *do not* just debate.
(5) Accept that in consultation you lose some arguments.

Decision	*Authors*	– Objective of output – Professional skills – Present method of operation
Consultation	*Authors*	– Alternative methods of operation
	Management services	– Parameters of acceptability of output
Decision	*Management services*	– Method of input
	Administration	– Method of output

FIGURE 9.1 Areas of decision and consultation; partial content – authors

Decision	*Typists*	– Choice of job – Relaxation – Colour scheme
Consultation	*Typists*	– Pay
	Unions	– Working conditions (noise)
	Personnel	– Redundancy pay
	Administration	– Working hours
	Management services	
Decision	*Management services*	– Office layout
	Administration	– Job content – Training

FIGURE 9.2 Areas of decision and consultation; partial content – typists

Decision	*Authors*	– Content of output – Objective of output
Consultation	*Authors*	– Output meeting objectives
	Management services	
	Administration	
	Typists	
Decision	*Management services*	– Typeface
	Administration	– Quality of paper
	Typists	– Keying method

FIGURE 9.3 Areas of decision and consultation; partial content – output

In conclusion it is well to remember the following quotation:

> It is a bad plan that admits of no modification (Publilius Syrus, Sententice No. 469).

TRAINING

In the past, trainers have often resorted to the methods of Skinner and Bloom. Their work was practically applied to training design projects in the USA in the 1960s, and in Britain in the early 1970s. This systems approach to training looks at training design as a cyclical process, each element in the cycle interacting with the others to modify them. Six elements make up the cycle:

(1) task analysis;
(2) specifying objectives;
(3) determining the target population;
(4) designing training;
(5) carrying out training;
(6) assessing the training effectiveness.

The approach has been fairly successful in almost every situation, and the investment by an organisation in building suitable packages for training has been cost-effective. However, the world of office equipment is in the melting pot. A word processing operator today could be an architect's assistant tomorrow, or programmer, or relegated to a simplistic task such as feeding a paper output device. The reason for these changes is that as technologies merge, and electronic mail, information retrieval, data processing, graphic design, voice response, voice input, etc. become available to existing equipment, they make the original task analysis incorrect. One could naturally prevent beneficial change (if you are in a position to monitor it) by retarding the rate at which you allow the use of facilities. The logic of preventing the use of advanced equipment in order for your training department to rebuild its obsolete training package may not, however, be too popular, either with staff, or with financial directors looking at profit margins.

> Chaos often breeds life, when order breeds habit (Henry James).

The use of high technology presents one other problem. As office equipment is still in its infancy, new developments are consistently changing the shape of the products. A cyclic change of between 5 and 7 years in hardware may well become normal practice. Obviously, manufacturers will encourage this by withdrawing maintenance contracts on existing equipment, and encouraging new sales by offering discounts and trade-ins.

In conclusion, the challenges to the trainers in this scenario of the training environment for office automation are as follows:

(1) use of cheap office equipment with high or low skills;
(2) constant change as equipment software is developed;
(3) cyclic hardware change;
(4) labour force being decentralised or even working from home;
(5) some tasks being allocated to self-employed staff.

AN APPROACH TO TRAINING IN THE AUTOMATED OFFICE

Step 1 – Define objective

When approaching any problem the 'objective' you wish to achieve at the end should be fiscally specified. If you are unable to make this form of numerically tangible comparison you will find it difficult, if not impossible, to monitor progress during the exercise, and to measure whether you were successful or not at its termination.

> The secret of success is constancy to purpose (Benjamin Disraeli).

Office automation objectives may concern fundamentals of a company's business, such as marketing, or administrative sub-routines such as word processing. These typical examples could be expressed as follows:

To reduce the cost of selling each motor vehicle
To reduce the cost of text production

One other important issue is to ensure that the organisation has only one objective and not several, diametrically opposed to each other.

Step 2 – Planning stage

Normally, the innovative management services officer or automation specialist will have selected equipment, and will now be asking the trainer to optimise on the staff available to use the equipment, and to convert organisational and procedural requirements into a training course.

As with most training approaches one needs to ensure adequate time has been allowed for the trainer to accommodate the request, and input to an implementation plan may prove profitable. Appendix C details a word processing and administrative support implementation plan.

> The success of most things depends upon knowing how long it will take to succeed (Montesquieu).

Step 3 – Instructional phase

Equipment manufacturers normally supply a short instructional course, often run by untrained demonstrators in an effort to sell their equipment quickly. This would normally be avoided, and a discount negotiated in exchange for carrying out your own training. The elements of a sound instructional phase are the following:

(1) identifying the basic instructions required for the normal work. A Pareto approach should be used here (i.e. what 20 per cent of instructions would allow 80 per cent of work to be completed?);
(2) using a building-block technique for imparting knowledge (i.e. learn to switch on machine, and demonstrate this knowledge, before being taught what to do next);
(3) self-paced lessons (preferably supplied one at a time as preceding lessons are mastered) – a sample lesson is reproduced in Appendix D;
(4) on real work (production must be seen to be worthwhile and profitable by learner);
(5) independent of teacher (the instructions must be simple enough to be self-supporting so that they could be used at home);
(6) selection based on success (any selection from a group of workers to use the equipment must be based on the student's ability to demonstrate success in learning on your course, and not on any previous performance). (See example Matrix for selection of word processing operators; Figure 9.4.)

Step 4 – Development

As the equipment comes to be used by staff and your development team gains experience, new cost-effective methods of operation will become apparent. This will require changes to original instructions given to the users of equipment. It is, therefore, advisable to subdivide assistance training in this area.

Surname	Christian name	Title	Quan.typed ESITs time	Orig. errors char's	Amend errors char's	Operational test
Baran	Anna	Mrs	434/90	54	40	23/25
Curry	Pamela	Mrs	408/90	150	137	21/25
Duckworth	Jean	Miss	469/78	93	73	23/25
Ginley	Eileen	Mrs	464/90	235	248	22/25
Haigh	Beverley	Mrs	469/90	241	201	24/25
Jackson	Jean	Miss	469/76	70	73	24/25
Kay	Hilary	Miss	462/90	41	31	24/25
Newman	Winifred	Mrs	416/90	30	32	23/25
Nicholson	Patricia	Miss	469/73	302	199	22/25
Padley	Barbara	Miss	469/90	83	109	22/25

FIGURE 9.4 Selection of word processing operators

(1) A reference manual of instructions, preferably both in the system in user-responsive form, and hard copy in decision-table style, or algorithms.
(2) A reminder guide to the instructional training course. (Perhaps in cartoon form, or a slide and tape presentation)

> When you know a thing to hold that you know it; and when you do not know a thing to allow that you do not know it; this is knowledge (Confucius).

(3) Instruction on procedures or specialised applications appearing on the input forms to the user.
(4) Occasional in-house seminars organised by yourselves or external consultants to motivate development, discussions, and pass on experience.

Step 5 – Change
The realisation that 'Change is not made without inconvenience, even from worse to better' (Samuel Johnson) may be somewhat alarming if you predict that every 5–7 years hardware and, possibly, revolutionary procedural changes, are going to be made. But if you can make staff aware by conferences, debates, in-house seminars, films, etc., that the only certainty is change, then a predicted cyclic change will become acceptable.

Step 6 – Repeat

With a change of hardware the process of OBJECTIVE, PLAN, INSTRUCTION, DEVELOPMENT, CHANGE, will be repeated.

ORGANISATIONAL IMPLICATIONS OF INTRODUCING WORD PROCESSING

At the City of Bradford Metropolitan Council the typists were grouped together in four typing pools, three in Jacobs Well and one in City Hall. There was generally no specialisation, and typists were called upon to do a number of miscellaneous tasks such as photocopying, reception work, collating, etc. The typists were located geographically near to some authors and although they regarded themselves as working exclusively for these, in theory, the typing pools were supposed to distribute the work evenly. An organisational structure is shown in Figure 9.5. The problems associated with a large reduction in staff and specialisation of certain functions of text production required the structure to be fundamentally changed. The organisational structure implemented is shown in Figure 9.6. The significant changes were as follows:

(1) non-typing work was allocated to an Administrative Support section;
(2) each building had only one geographic location for typists;
(3) work was even more specialised through a further subdivision of the typing process;
(4) management services provided technical support.

THE 'U TURN' SHARED-LOGIC SYSTEM OF WORD PROCESSING

The attraction of specialisation was that it would optimise the use of the

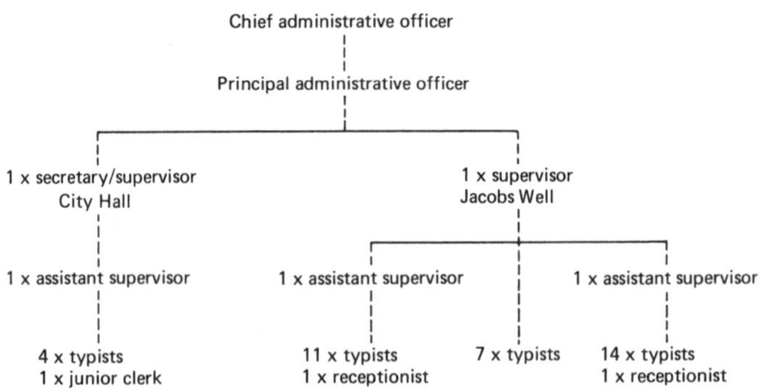

FIGURE 9.5 Organisational structure – 1975

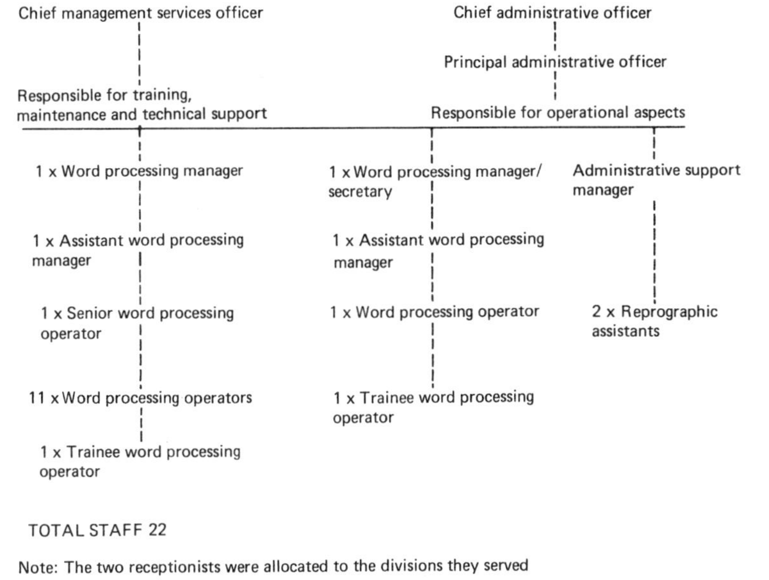

FIGURE 9.6 Organisational structure – 1978

equipment, and I therefore adopted the 'U Turn' shared-logic system of word processing.

The applications used by authors were re-designed and the stationery changed as necessary; subsequently these changes were implemented and procedures introduced for future amendments.

Essentially, instructions for completion of applications (typing jobs) appeared on various input forms, some of the information being pre-printed, and some being written on by the authors.

A typical operational sequence, using the appropriate input form, is detailed in Figure 9.7. The diagram shown in Figure 9.8 represents the layout of the two word processing centres. Both centres conformed to a system of paper movement similar to a 'U turn', a figurative example of which may be seen in Figure 9.9. Appendix D gives the equipment specification.

THE ARGUMENT AGAINST SPECIALISATION

Many observers concluded that, though the changes, once made, would achieve their objective, the resulting effects in staff morale would dissipate the savings.

Author	Word processing manager	Word processing assistant manager	Word processing operator	Word processing trainee
1. Completes typing input form and sends to word processing centre.				
				2. Time stamps incoming input forms.
				3. Places incoming typing in Manager's 'IN' tray.
	4. Divides input forms into batches according to paper types.			
	5. Allocates a document number.			
	6. Allocates a batch of work to an operator according to her daily work programme.			
			7. Types document and sends for printing according to paper type. Places draft in 'OUT' tray.	
				8. Collects input forms from 'OUT' trays and places in appropriate paper type pending tray.
		9. Instructs trainee in loading of appropriate paper to printers.		
				10. Loads printer for next batch of documents.
		11. Allocates the appropriate paper type to the printer.		
				12. Unloads printed documents from printer.
				13. Trims documents on burster and collates them.
				14. Collates printed documents with input forms from pending trays.
				15. Time-stamps outgoing work.
				16. Places work in 'OUT' pigeon-holes.

FIGURE 9.7 A typical operational sequence

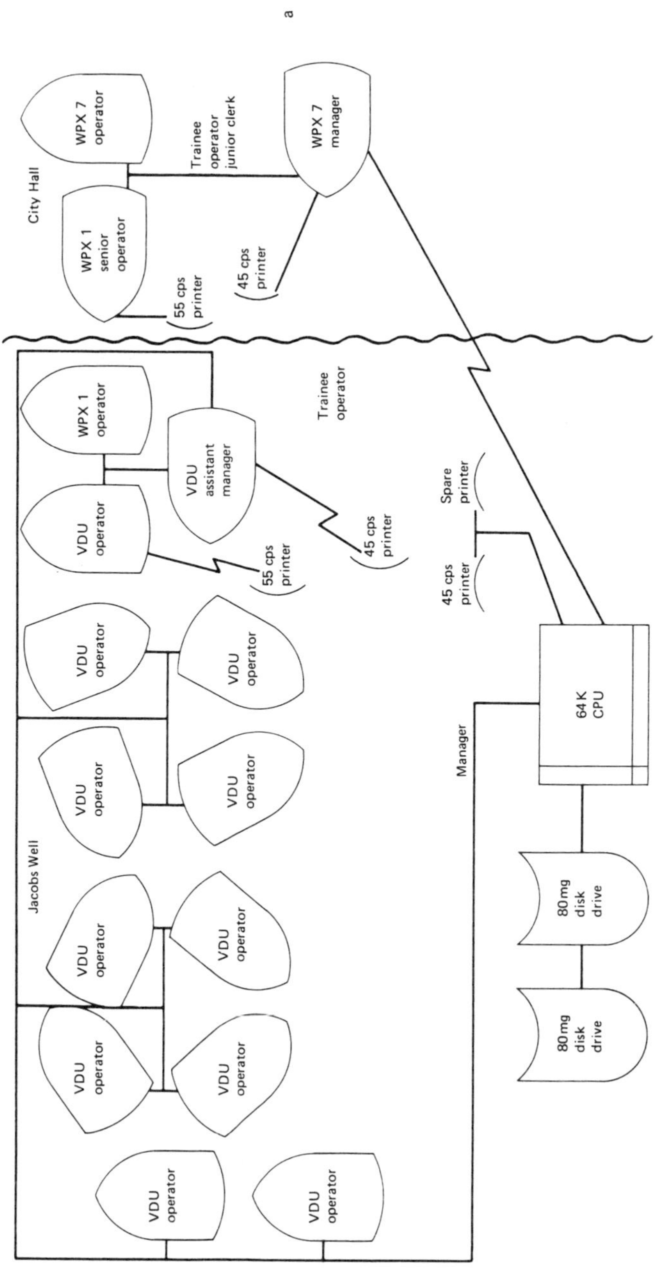

FIGURE 9.8 Word processing centre layout – Bradford

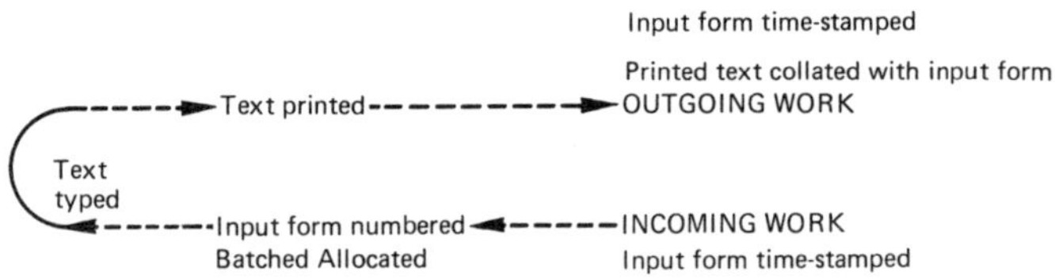

FIGURE 9.9 Diagrammatic example of 'U turn'

Organisational theorists like Hertzberg and Maslow argue that one of the prime motivations in completing the task is job satisfaction. Woodward and others argue that job satisfaction is essentially derived from the completion of a job from its start to its end. Observers therefore concluded that the application of word processing to a secretarial job subdivides a complete job, the typing and clerical operations being allocated to two quite separate people; even separate structures.

The effects of the specialisation would therefore mean a drop in morale with the following potential by-products:

(1) high labour turnover;
(2) high error rate;
(3) poor productivity.

BASIC CORRESPONDENCE SEQUENCE

The flaw with arguments concerning overt specialisation of typists is that the observer has assumed that the role carried out by secretarial services is a complete job, when in fact it is only a portion of a job. If one examines the basic correspondence sequence below, one will see that there are six distinct parts. Only the fourth and fifth parts are normally completed by the typist, i.e., typing the letter, and seeing that it corresponds with the author's composition:

(1) correspondent initiates a reply;
(2) author decides on reply;
(3) author composes reply (letter);
(4) typist types reply (letter);
(5) author/supervisor checks letter;
(6) letter answers correspondent.

The skills which enable a typist to justify her existence are the receipt of text (i.e. shorthand, audio) and the quick and accurate keying-in of the text on her keyboard.

The second and third parts of the correspondence sequence are unconnected with the secretary's skills and relate solely to the author, although standard letters or documents at least allow the elimination of the composition of a response which has been made previously.

It may, then, be concluded that, in order to complete numbers (2)–(5) in the correspondence sequence, the author must also be the typist. It is therefore a prerequisite of the function of a typist to participate in only a portion of the complete correspondence sequence. So, any increased job satisfaction will only be derived from the typist identifying with her specialist role.

RESULTS OF SPECIALISATION

Now to the savings and improvements in efficiency: one indisputable fact stood out – where 44 members of staff sat before, only 22 remained and, what is more, the workload had increased to the tune of 19 per cent. Taking everything into account (wages, overheads, equipment costs, accommodation, development and installation costs spread over 7 years), the annual savings approximated to £58,000–59,000.

As to the level and standard of service before and after? – Well, I know of only three quantifiable criteria which enable an objective comparison to be made. These are: turnround time, error rate, and quality of print.

Unfortunately, there was no data available in respect of turnround times before the word processing system was installed. Accordingly, all that could be done was to ask the typing pool supervisors what their target times for each category of work had been.

With the installation of the word processing system we introduced an arrangement for the time and date stamping of each item of work – both in and out. A comparison of the two sets of turnround times is set in Table 9.1. Comparing the actual number of errors made, both before and after the word processing system was installed, was relatively straightforward; *before* installation we simply counted the number of errors during the typing survey – maintaining a regular record of errors *after* the installation became operational was a normal part of operating procedure. The 'problem' lay in relating the actual number of errors to some base which did not give an unfair advantage to word processing. (The system produces far more typed output per typist for the same effort, by using standard paragraphs, by 'recalling' pages and portions

TABLE 9.1 *Comparison of turn-around times (hours)*

Category of work	*1975 (Before)*	*1978 (After)*
Corrections and amendments	4	3.27
Letters and Memoranda	14	7.59
Labels	37	9.19
Forms	14	11.79
Reports	37	18.43

of text, etc.) Accordingly, I related the actual number of errors to the number of keys indexed by the operator, to produce the following comparison:

Error rate per keys indexed:	1975 (Before)	1978 (After)
	1:59	1:367

Finally, the quality on a QUME printer (the type used in the word processing centre at Bradford Met.) was much higher than that of a conventional typewriter. This was because the printer varies the intensity of pressure for each character type.

Naturally not all authors will be happy with change. Some will still prefer their own 'go-fer' – you know – the type of 'office wife' who goes for their coffee and shopping, and, incidentally, happens to type as well. A systematic change-over to a standardised service will remove these unofficial services and some authors will crib. However, at Bradford, a survey undertaken independently by the administrative officers demonstrated that, generally, authors preferred automated word processing.

AUTHORS SURVEY

75.3 per cent authors thought the system operated as well or better than previously;

69.4 per cent authors thought the quality of work was as good as or better than previously;

51.6 per cent authors thought the turnround time was as good as or better than previously.

To what avail, though, such sizeable savings and an immediate

TABLE 9.2 *Attitude survey*

Within which environment would you prefer to work?	Word Processing Centre	Previous Typing Pool	Another Job
	12	2	Nil
Which type of equipment would you prefer to operate?	Wordplex7	Electric typewriter	Manual typewriter
	14	Nil	Nil
Which type of post do you find more challenging and interesting?	Word processing operator	Copy, audio or shorthand typist	Secretary
	13	Nil	Nil

NOTE: 14 staff returned survey forms. One person did not answer the last question.

improvement in efficiency, if labour turnover rockets and the typists' morale plummets?

One year after installation not one operator had left, whereas turnover was previously 35 per cent, and an attitude survey carried out (from which the results are shown in Table 9.2) suggested that the operators were much more satisfied than they had been previously. Perhaps for typists we need a new behavioural theory? How about 'specialisation groups'?

APPENDIX A

EQUIPMENT SPECIFICATIONS AES WORDPLEX 7

One central processing unit

64K high-speed semiconductor memory (using 4K RAM technology)
Dual processors interleave memory cycles
Inline hardware diagnostics
4.8 Mbit/s direct memory access facilities interleave memory cycles with processors
Processor accesses up to 64 I/O channels
Multiplexer permitting 32 full-duplex asynchronous communication channels to be operated at 9.6 kbit/s
16 384–65 536 byte addressable memory
Memory cycle time at 550 ns
Transfer rate of up to 4 800 000 byte through DMA to/from memory

Power	240 V AC, 50 Hz
Weight	181 Kg (400 lb)
Maximum Floor Loading	8.8 kg/cm^2 (125 lb/in^2)
Size	34.0 in × 48.0 in × 18.0 in (height × width × depth)

Keyboards [one per VDU]

Key switches	Solid State Hall effect type MTBF = 1 M h MTBF = 20 G operations

16VDUs

Window size	12 in diagonal 24 lines by 80 characters
Page size	83 lines by 128 characters
Matrix	7 × 9 dot matrix, sans seriff
Colour	Green or grey
Brightness	Variable
Intelligence	12 VDUs of 8k and four VDUs of 32K
Power	240V AC 50 Hz
Weight	55 lb
Size	12.4 in × 18.5 in × 18.6 in (height × width × depth)

Two storage module disc drives

Capacity	80 Mbyte
Basic unit	Single spindle drives using five-high disc packs, CDC 9877 or equivalent
Start time	22s
Stop time	90s
Access time	10 ms (1 cylinder movement) 30 ms (average movement) 55 ms (822 cylinder movement)

Keyboard enclosure	High impact, flame retardent ABS plastic
Key depression interval	0.4 ms
Total number of keys	83
Keyboard control keys	4
Cursor control keys	7
Software assigned keys	2
Editing keys	25
Shiftlock Indication	LED
Delay before repeat	0.6 s
Slow repeat rate	15 character/s
Fast repeat rate	30 character/s
Flexible cable length	5 ft
Weight	5 lb
Power	Supplied by VDU
Size	3.3 in × 18.5 in × 8.0 in (height × width × depth)

Four dual flexible disc drives

Track-to-track access	
Time	10 ms
Size	10 in × 20 in × 13 in

Disc pack

Usable surfaces	6 (5 data, 1 servo)
Diameter	14 in
Coating	Magnetic oxide
Size	10.0 in × 200 in × 13.0 in (height × width × depth)
Pack-rotation period	16.7 ms at 3 600 rev/min
Average rotation latency	8.35 ms at 3 600 rev/min
Cylinder per disc pack	822
Tracks per cylinder	5 data tracks 1 servo track
Tracks per pack	4 110 data tracks
Track capacity	20 160 byte/10 080 word
Cylinder capacity	100 800 byte/50 400 word
Pack capacity	82.4 Mbytes/41.4 Mword
Data-transfer rate	3600 rev/min: 1.209 Mbyte/5 9.67 Mbit/s
Recording method	Modified frequency modulation
Size	34.0 in × 19.0 in × 34.0 in (height × width × depth)

Five printers

Printing speed	45 character/s bidirectional
Printing method	Daisy wheel and servo hammer
Print Wheel	96 character ASCII daisy wheel, interchangeable, plastic
Paper Handling	Bottom-fed continuous stationery, sprocket held
Width	15 in
Ribbon	Cartridge
Controls	Horizontal forms positioning, vertical forms positioning, forms thickness, ribbon advance
Size	7.0 in × 23.5 in × 14.5 in (height × width × depth)

APPENDIX B

STRATEGY FOR THE INTRODUCTION OF WORD PROCESSING (WP) AND OTHER OFFICE AUTOMATION

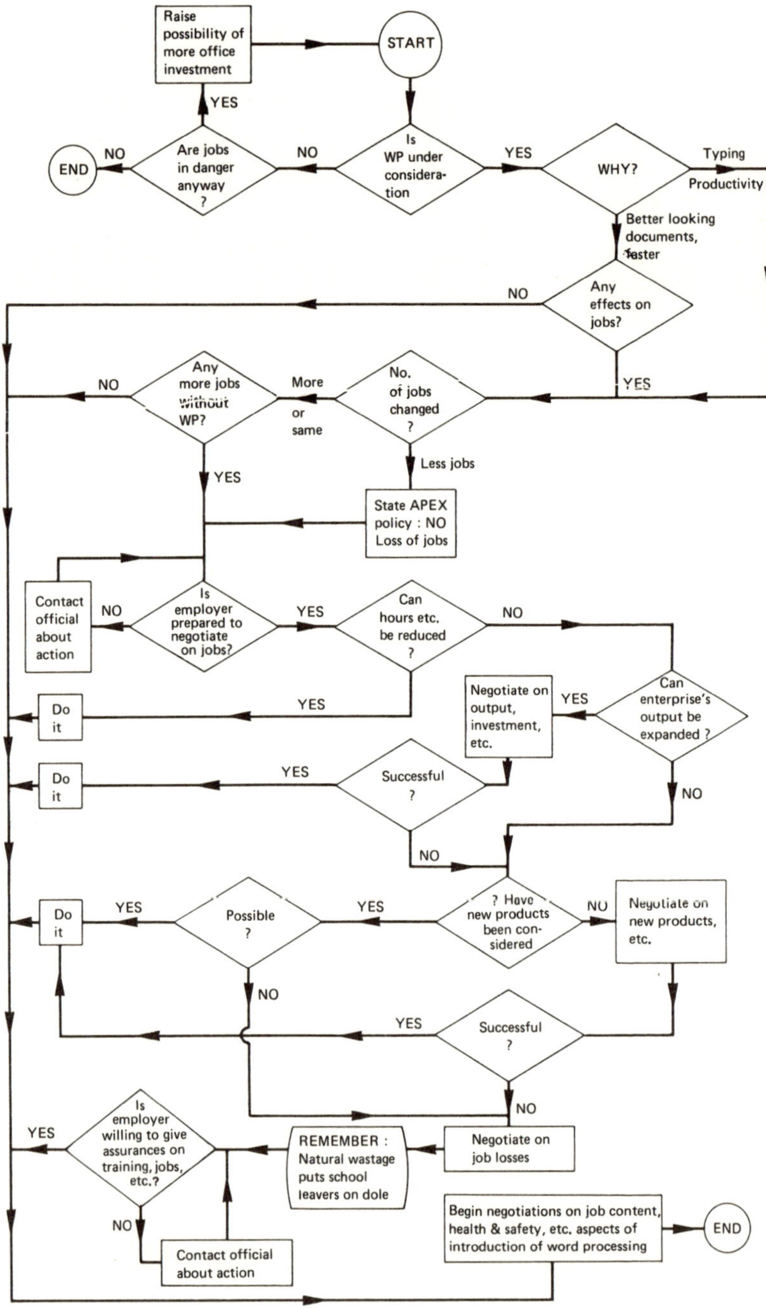

APPENDIX C

WORD PROCESSING AND ADMINISTRATIVE SUPPORT DETAILS OF IMPLEMENTATION

Months after implementation go-ahead	*Month* *1 2 3 4 5 6 7 8 9 10 11 12 13 14 15 16 17 18*
ACTIVITIES	
1. Career and personnel plans and policies	
(a) Staffing plan	
(b) Job descriptions	—
(c) Career ladder	
(d) Pay plan	—
(e) Develop tests	—
(f) Interviewing, testing and selection	—
Management and supervisors	
Administrative support staff and others	— — — ——
(g) Personnel manual update	————
(h) Review and update	—
2. Management and staff orientation	
(a) Prepare materials	
(b) Give presentations	—
3. Training	
(a) Plan prepared	
(b) Formal materials prepared	—
(c) Site visits and outside instructional classes for:	—

APPENDIX C (*continued*)

Months after implementation go-ahead	*Month* 1 2 3 4 5 6 7 8 9 10 11 12 13 14 15 16 17 18
Supervisors	—— —— —
Administrative support manager	——
Word processing manager	———
(d) Inhouse development training programme for:	
Authors	——————
Administrative support staff	——————
Word processing operators	————————
(e) Review and update	—
(f) Change prepared	
4. Facilities and equipment	
(a) Discuss floor plan, decor, noise and furniture with staff and agree layout	——
(b) Dictation equipment review and selection	—
(c) Text editing equipment ordering	—
(d) Copying equipment ordering	
(e) Accessory supplies ordering	—
(f) Initial equipment received and in facilities and equipment	——
(g) Equipment review and changes	

APPENDIX C (*continued*)

Months after implementation go-ahead	*Month* 1 2 3 4 5 6 7 8 9 10 11 12 13 14 15 16 17 18
5. Forms and procedures	
(a) General procedures	
Principal to manual	——
Initial reporting procedures to principals and management	——
Initial style book	——
Glossary	——
(b) Administrative support centres	
Managers' manual	——
Assistants' manual	——
Miscellaneous forms review and design	——
Messenger's mail procedures	——
Phone follow-up procedures	——
Filing systems design and implementation	——
New special projects for administrative secretaries	——
(c) Word processing centre	
Manager's manual	——
Operator's manual	——
Quality control	

APPENDIX C (*continued*)

Months after implementation go-ahead	*Month* *1 2 3 4 5 6 7 8 9 10 11 12 13 14 15 16 17 18*
system	
Production control system	
Operator standards system	
Applications improvements	
(d) Word processing reviews	
Administration	
Developments	
Education	
Finance	
Social services	

APPENDIX D

WORD PROCESSING TRAINING COURSE – LESSON ONE

This first lesson aims to familiarise you with the switching on and off procedures, and with how to 'get into' and 'out of' the system.

1. HOW TO SWITCH ON AND GET INTO THE SYSTEM

1.1 *THERE IS A BLACK SLIDING SWITCH AT THE RIGHT HAND SIDE OF YOUR VISUAL DISPLAY UNIT (VDU). SLIDE THIS AWAY FROM YOU TO THE 'ON' POSITION*

1.2 *DEPRESS THE KEY ON THE TOP ROW MARKED 'STOP CONT'. WAIT A FEW SECONDS*

Your VDU is now displaying the question 'WHO ARE YOU?' This is what you will see first thing each morning. The flashing light is called the cursor. This indicates the position at which your typing will appear when you begin to type.

The brightness of the screen may be adjusted to suit you. The black knob at the right hand side of the VDU controls this. Turn it either way until the brightness is suitable for you.

1.3 *NOTICE THAT THE LOCK KEY IS DISPLAYING A SMALL RED LIGHT. THIS IS THE SHIFT LOCK INDICATOR. AFTER SWITCHING ON THE VDU THIS LIGHT WILL ALWAYS BE ON. TO REMOVE IT, PRESS THE SHIFT KEY.*

1.4 *TYPE YOUR OWN CHRISTIAN NAME WHICH WE HAVE RECORDED ON THE SYSTEM AS ANN*

1.5 *PRESS THE 'RETURN' KEY*

The screen will now display 'ENTER PASS CODE'. When the pass code is typed it will *NOT* be displayed on the screen.

1.6 *TYPE YOUR OWN PASS CODE WHICH IS 'MINE'*

The screen will now be displaying 'ENTER COMMAND'. This means that the system is awaiting your instructions. You are now into the system and are in a position to begin work, i.e. typing, corrections, printing, etc.

2. HOW TO GET OUT OF THE SYSTEM AND SWITCH OFF

At this stage it is not intended that you should do any typing. The command you will now give is the one used to get out of the system, normally given at the end of a working day or period.

2.1 *TYPE THE LETTER B.*

The screen displays 'BYE'.

2.2 *PRESS THE RETURN KEY.*

You are now back to the 'WHO ARE YOU' stage.

2.3 *SWITCH OFF YOUR VDU BY SLIDING THE BLACK SWITCH AT THE RIGHT-HAND SIDE OF THE SCREEN TOWARDS YOU.*

You should now repeat this switching on and off procedure until given further instructions.

10 Microelectronics for Management in Service Industries

PETER G. HARDIE-BICK

INTRODUCTION

In the early part of the nineteenth century steam began to replace horse,- wind- and water-power. Early installations found considerable economy in scale and large steam-powered machines grew to be commonplace. This tendency for large machines to have an advantage persisted until the early years of the twentieth century when the internal combustion engine became a practical proposition for powering motor cars. Subsequently, power units have become smaller still until nowadays no one feels it to be strange for individuals to possess very high-powered machines for their personal use. Just think about the consequences of using a chariot drawn by eighty horses to go to the office in the morning! The personal mobility and freedom of choice which accompanied this development of high-powered machines, and their virtually unrestricted use by individual human beings, has much to do with the way our civilisation has developed in the twentieth century. Even allowing for the effects of relatively higher fuel costs this freedom of mobility and the ability to command vast amounts of power, by a movement of one's fingertips, will continue to be a dominant feature for the foreseeable future.

Computing, using electronic machines, has followed a comparable pattern which has evolved over the years since 1945 when the first really effective machines were devised in the UK. Since then, the first three decades were to follow the line of increasing emphasis on economies of scale until the early 1970s. These economies of scale have been genuine and the use of computers has become widespread mainly because improvements in telecommunications allowed individuals to use them through normal telephone links connected to a relatively cheap and portable keyboard terminal. Networks of these links have grown to a point where computing is rather like an electrical power supply. One may be scarcely conscious, there is no need to be, of where in the world our computing is being done. This may even apply to data storage. This is hardly ever the cheapest way of carrying out routine tasks, however, because there is always a cost arising from the communications element and the management and operating systems essential to large computers, which are available to many users simultaneously, are costly to the point of eliminating the advantages of scale.

The influence of increasing economies of scale in data processing has also had its effect on organisations. It had become commonplace, by the early 1970s, to have quite large single computers serving a substantial enterprise. The effect of this channelling of data processing into and out of a single location has had disadvantages as well as advantages. But as long as there has been a significant advantage in capital costs, the less tangible factors which have been disadvantageous have been largely ignored. These intangible negative factors are usually found in any centralised bureaucracy, but for some reason it comes as a surprise that such an otherwise up to date concept could rapidly change our way of thinking about problems into a kind of modern byzantinism. Typically we find the following features:

(1) Fears about the security of data give rise to complex systems designed to limit access. These almost always turn out to be vulnerable despite the care that goes into their design.
(2) Systems design and programming become skills used by a highly trained elite brought up to restrict access to the machine to those sufficiently qualified to use it.
(3) Users begin to wish to escape from the straight jacket imposed by the very long time-scales which begin to emerge both for the introduction of new procedures and for the updating or modification of existing programs.

Developments in the fourth decade of electronic data processing were to provide some relief in two of these three areas of concern. First, the progressive reduction in the size of the circuitry required by computers by the adoption of transistors to replace thermionic valves and then by the remarkable evolution of smaller and smaller transister circuits led to the development of minicomputers. This allowed small, but very powerful, machines to be dedicated to single applications. Thus users found it possible to obtain the facilities they needed from machines which were both under their own control and close at hand. Secondly, although the minicomputers were not as flexible from either an operating point of view or in their ability to use programs written in the commoner versions of high level languages such as BASIC, FORTRAN or COBOL, they operated well in their own particular versions or dialects of languages.

The languages used by the minicomputers were designed for rapid assimilation by a new breed of users. These new users were unlikely to be too familiar with larger-scale computing although some were, to their advantage. The storage of data and its protection and recovery has sometimes needed considerable skill. But this has been regarded as less of a disadvantage, for numerous applications, than the difficulties arising from becoming tied to larger-scale computers when local requirements pointed to a need for security and rapid response to change.

This reversal of the arguments in favour of large scales of operation in computing brought a reaction which was to lead to a high rate of development of new uses for the larger types of computer which had been so rapidly evolved over the previous decade. Indeed it should be observed that the older technology, as in the case of sailing ships when they were challenged by steam, reached hitherto unknown heights of efficiency through competition with the new. The vast databanks developed in the early 1970s by several companies in the USA, and closely followed by agencies of the EEC, would never have emerged had it not been for the cheapness of the data storage available through these means in a state of high competitiveness with all other forms of data processing. Furthermore, although the time-sharing facilities, so abundant in the 1970s were made available mainly through the sophisticated telecommunications techniques which had been developed, they were also dependent on the availability of supplies of data processing capacity provided on demand from the large 'time-sharing' computers. These had by then been installed at many locations through the world.

But the state of equilibrium afforded by this relatively stable interregnum period, when the competition, if it can be called such, between the large machines and the minis was at its height, was greatly disturbed by the rapid development of the microchip.

The microchip is not really a revolutionary device; in fact, computer components have been reducing in size progressively ever since they were first invented. It must be admitted, however, that the state of the art has been advancing remarkably in comparison with the technology available only a short time ago. Silicon technology now allows thousands of components to be incorporated in a small stack of laminations a few millimetres square. Not surprisingly, therefore, one can build a powerful computing device by this means. Linking a few microchips together can produce something very like the minicomputer which itself challenged the older styles of computer less than ten years previously. For the manager, this new development has the most profound implications.

A NEW RANGE OF OPPORTUNITIES

The possibility of holding in one's hand a few components, intrinsically inexpensive, which can carry out many different instructions at the speed of light, is a remarkable development. The services they can provide are not so very different from the large data processing installations so familiar in retrospect. And yet it may not be so easy, initially, to see how such devices can fit into the pattern of everyday life by providing services of this kind. It may be helpful at this stage to speculate a little on the kinds of service which managers might most appreciate and in which perhaps certain useful features may become more easily available by such means.

INFORMATION SERVICES

Nowadays managers are only too familiar with the deluge of information which arrives each day. It is difficult to take the initiative about information that arrives like this. Almost inevitably it does not arrive when we need it and then it gets buried unless we have a good filing system to recover it later on. It would be better to file it right away and recover it at exactly the time and place it is required. This is what an electronic filing system can do in response to a management request. At a personal level this might be a bank statement or a list of cut-price refrigerators available from local hardware stores. It might be the sales figures for the last week, up to yesterday rather than last Friday; or the state of the depot stocks and impending receipts and despatches for the next week – not the next calendar week, but the next seven days from now. Information of this kind produced regularly would increase the deluge beyond toleration; but on request at a particular time it might be vital to the success of, say, the launch of a new product where both one's own and competitive promotions are in operation. The ability to respond flexibly to changing requirements is clearly a most desirable characteristic. Taking new data and presenting it in a particular form rather than any other may mean the difference between a manager being able to support his hunch with hard data or not being able to do so.

SPECIAL COMPUTATIONS

One of the features, which has accompanied the introduction of computers, has been the elimination of the useful jobbing shop provided by comptometers. Often, additional calculations are needed to provide special information about, say, the performance of a product in the market under special conditions, or perhaps the optimal policy for reordering supplies of a particular item when a price increase is about to take place. Such information is usually needed either at very short notice or not at all. There is no easy way of ensuring that it can be provided through large-scale computing installations other than by either anticipating the need and meeting it through a variant of standard procedure, or by writing a special programme to make an *ad hoc* report through the channels which have been set up for the purpose. This can lead to very complicated programming. Once again, managers need to be able to ask questions and to get answers in a reasonable time if they are to fulfil their role effectively. Therefore, an additional series of calculations to make a special report should be easy to produce and so should further calculations arising from questions posed by initial reports.

If hard information of this kind could be produced on request, it would transform the role that such information has come to play in the decision-making process. Paralysis through too much analysis is a feature of many a special exercise, but this is only because the basic reporting system is often too rigid to be used or interpreted in a wide variety of ways. As a result managers have had to rely more and more on the 'soft' intuitive type of information. The terms in which soft data is presented are precisely relevant to the point in question. Indeed qualitative information raises the most vital question. A flexible and rapidly responding service to provide the hard data needed to back hunches of this kind would be invaluable to the manager in an operating situation.

AVOIDING OVER-ORDERING

The commonly encountered phenomenon of the rapidly vanishing shortage in a distribution system has frustrated many a wholesaler and brought some manufacturers to their knees. There are many different examples but perhaps typically, a new product may have been launched with some initial successes such that stocks in the field became quickly exhausted. The question, which requires a rapid answer to avoid attracting the avid attention of competitors, is how to establish a satisfactory delivery plan to retailers without overstretching the manufacturers, who may in any case have a restricted capacity. When this has to be done under pressure with customers' orders piling up at an unprecedented rate, it is not surprising that over-ordering begins to occur throughout the system. This begins with the retailer who may feel that a bit of optimism in his ordering will not be to his disadvantage when supplies are short. The wholesaler himself may feel he has to pass on orders for quantities to meet his customers' requirements plus some additional stock cover for himself. He should feel able to meet at least some orders from stock if he is to provide a service to his retailer customers. The manufacturer is in a more difficult position. If he quotes realistic delivery dates he may lose the orders but if he increases capacity to meet them he risks a great deal, especially because his working capital requirements may have to be increased too. Most wholesalers and manufacturers who have even once been in this position are quick to recognise the danger of over-reacting to what appears to be a very large increase in consumer demand for their products. The intuition of the warehouse manager will say, at the most appropriate time; 'this is too much, this shop has just ordered a year and a half's supply of stunting kites'. But they have been successful in other places and this type of kite may have increased the size of the market. His intuition needs some quantitative backup to support the hunch with hard data. After all, orders must be placed. The question of how much is really quite difficult to answer. There has been too short a history of sales for this particular product to provide anything but a rather optimistic prognosis.

One way to make an estimate under these circumstances is to say 'what if these orders were realistic; would they appear reasonable in the light of the sales figures I have for toys of this kind in the past few years'. At this stage some hard data would be most helpful. But data may not be available in a form that is suitable for the purpose. It will almost certainly be in terms of sales by value which will include an inflation component nowadays, and overall turnover figures may be affected by a number of other extraneous circumstances such as the ambiguity of the classification of toys and sports equipment as in a craze for skateboards in which the classification has never been clearly defined.

Sampling might produce some useful pointers, but by sampling one does not consider significant random sampling as a practical way to establish a trend within the limits of accuracy provided by the sample. On the contrary, the sample here would be any unit in which conditions allow it to perform reasonably and understandably. Thus, in a situation like the one described, it might be useful to examine the sales performance in a unit which had not run out of stock. Finding such retailing units in a relatively large population would entail asking a series of questions about stock levels, orders, sales, other retailers in the area and any special information about the retailer concerned. Thus one might be able to select, from a relatively large population, those retailing units on which the most reliable estimate of demand could be based.

It is likely that a feeling of great shortage in a distribution system will be caused by a shortfall in supply of a few per cent. 'Soft' information from the market place is more likely to provide reliable data about the nature and extent of the shortage than the most commonly available 'hard' data which is in the form of orders received from the next level in the system. If 'hard' data could be found to test out hypotheses about demand by selecting samples intelligently from the field, one could better expect to be able to quantify the hunches obtained from soft data.

DECIDING HOW MUCH TO ORDER

Once a level of demand has been established, how much to order is a question which is a challenging puzzle to managers. In the case of very high demand items, it may be most convenient to base order quantities on the consumption between reviews which are themselves based on a pre-set pattern of deliveries. But for items which do not fall into this category many factors may have an influence on the quantity to be ordered. If the decision to order is to be based on an appraisal of the stock level, probably triggered by it, the quantity required may be determined by economic factors such as cost, the extent of free credit, the value of stock held or volume of storage space required.

But cost in this case is not just the interest on the cash committed through an order. It is also a function of the cost of storage and the rate of spoilage in addition to the costs associated with ordering and receiving the item. Furthermore, there are several limiting factors, such as the minimum size of order that a manufacturer or distributor will accept for an item or a group of items and, vitally for those placing the order, the maximum length of time an order can be allowed to cover. A year's supply of a particular item might be regarded as an upper limit unless unusual influences were at work.

When such a large number of factors affect decisions which must be made frequently, there is a tendency to fall back on the 'soft' data which may just be past experience in similar situations, or drastic simplification may be needed to establish a rule of thumb. Thus classification of items into broad categories, based on demand and value, has provided many wholesalers and retailers with a pragmatic set of rules for re-ordering.

Other factors may be temporarily influential in deciding how much to order. A price increase may be imminent or a special discount available for orders in excess of a certain amount. Perhaps credit is short and orders should be set at the lowest possible level even if this incurs some cost disadvantage. When added to the already quite complex list of variables, these and perhaps other factors contribute to making the re-ordering process so complex a problem to rationalise fully that it has been rare to find an operational system that does not depend mainly on the interpretation of 'soft' data by the line manager concerned.

However, there is little doubt that in such systems the decisions made are not always the most economic which could be achieved in the circumstances. In using soft data people tend to under-react to small changes and perhaps also to over-react to large ones. A rational system based on quantification of all the measurable factors would provide better results if both hard and soft data could be employed in a judicial manner without overburdening the manager with detail.

CAN WE AFFORD TO OFFER A DISCOUNT?

Trading terms have become so much a part of the overall package offered to customers that it may be found that the standard price is rarely charged. There is no doubt that certain classes of customers cost more than others to supply. In addition, there is pressure on most distributors to offer goods for sale at nominally the same cost in different parts of the country where widely differing transport and storage charges will be incurred. Under such circumstances it is hardly surprising that costs may be suddenly increased by a change in the pattern of demand for, say, ice cream when summer comes and the population moves out of the cities into a number of more widespread centres. Customers have their own idiosyncratic approach which they may often partly impose on suppliers to their own advantage.

It is frequently the customers who are most successful at this; of course raising suppliers' costs, who are also the most tireless in negotiation of other terms of trade. Altogether the trade-terms package for an individual customer may turn out to be very complex indeed and, both in the course of negotiation and afterwards in practice, it may be extremely difficult to calculate the profit which will accrue to the supplier. Indeed when companies actually go to the trouble of making this calculation about their arrangements with major customers, it usually turns out that profitability varies considerably between one customer and another.

This pattern may not necessarily follow rational lines in which the most highly prized customer accounts might be expected to provide the highest net income to the company. Hard data of the kind most commonly available, about volumes and net margins per product group, are of relatively little help in this situation and management frequently has to fall back on hunch, particularly during negotiation. Better hard data about particular accounts' net profitability is difficult to make available, but when it can be provided, the improved balance between hard and soft data assures support to the manager, especially when he is negotiating. Yet again the variety of possible arrangements with customers is very wide indeed and providing hard data in advance to cover even a range of possibilities would give a deluge of information which would be most unwelcome. Experience tends to show that soft data about individual customers, in providing an assessment of the kind of deal that they are most anxious to achieve, will offer the best means of determining how terms of trade can be arranged at minimum total cost to the supplier.

A BALANCE BETWEEN HARD AND SOFT INFORMATION

These three examples of typical problems in the distribution area serve to illustrate the nature of the tasks facing the manager as well as the means at his disposal to carry them out. In every case intuitive or soft data plays an important part and managers are often called upon to 'use their heads' instead of concentrating solely on the hard, quantitative data which, though inadequate, may be immediately available.

It is paradoxical that most managers have found that the arrival of computers has tended to isolate them from the production of figure data.

This has become, in some cases, a factory process with specialist functions, concern for standardisation and so on. Managers have been asked to forecast their requirements for information, sometimes years ahead, so that the processes of systems analysis, design, programming and computerisation can be scheduled in the appropriate departments.

It would not be altogether surprising if managers were to become more dependent, through this experience, on their intuition and the interpretation of soft data about situations arising from day-to-day operations.

HOW MANAGERS THINK

Methods of problem-solving are almost as numerous as the managers using them. However, it may be possible to classify these methods in a way helpful to the development of some of the hypotheses which are evolving about the manager's role in the coming era of the micro-processer.

The problems which occur in business are not at all uniform in any of their important aspects. Solutions may be simply calculated or perhaps may require the full range of the intellect to even comprehend where they might lie. Generally, it is conventional wisdom to say that problems involving mainly calculated solutions are simpler than those involving other rational processes and perhaps emotional elements as well. Managers employ their own faculties in different degrees in solving their problems but their approach has been conditioned by their background and training as well as by their current role. The emphasis placed on hard and soft data by managers playing different roles can be tentatively illustrated by Figure 10.1.

It is not surprising that those most comfortable with figures often seek roles in analysis and planning whilst those whose inspiration comes from more qualitative aspects of life, where intuition also plays a large part, seek roles where judgement is essential to the performance of their work.

Evidence about this type of orientation in managers is not at all hard to find. Furthermore, it is becoming clear on the evidence of research into the activities of different parts of the brain that the left hemisphere concerns itself with the strictly rational processes which can be considered to include calculations, deductive logical operations, language and writing. The right hemisphere is held to be concerned with feelings, shapes and inductive processes which depend upon intuitive rather than rational thought. The spectrum of management roles in the

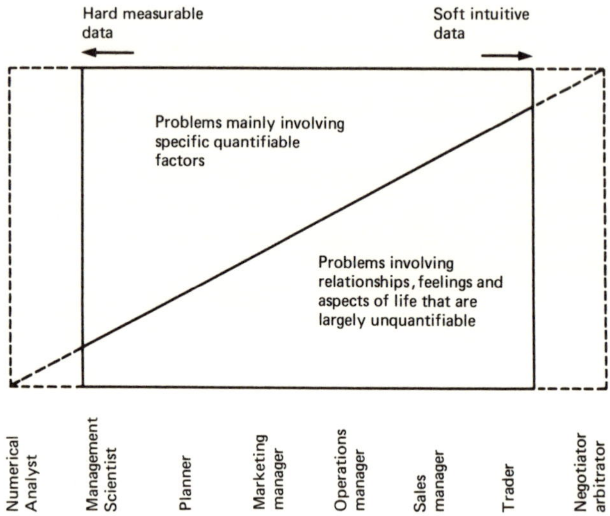

FIGURE 10.1 The manager's role and the relative emphasis on hard and soft data

diagram may be explained partly by the hypothesis that numerical analysts and planners are left-hand hemisphere thinkers whilst successful sales managers and traders have better developed right-hand hemispheres.

The thought that most computer systems seem to have been evolved by people who are left-hemisphere thinkers would explain their very mechanistic and in many cases rather rigid framework. Typical of this approach are the procedures often encountered in using computer systems with passwords, detailed procedures and forms which are absolutely essential to the successful operation of the system. They are, however, an effective blocking mechanism for the intuitive thinker who may wish to test a hunch about a situation which is developing. All he wants to do is to ask some 'what if' type questions based on a hypothesis of what may be about to happen. Usually, hard data is too difficult to manipulate in time because of programme or hardware architecture. So the natural consequence of an absence of hard data is to fall back onto soft data coupled with intuition to bring to bear a number of inductively linked arguments. For example, a question like 'why were sales low in Manchester last week?' might provoke an answer like 'we cannot provide statistics for Manchester because it is served by two depots which also serve other areas outside the town', from a planning section who were completely dependent on their computer records. So the sales manager would marshall all his sources of information which would give him a few pointers like:

(a) the weather in Manchester was awful last week;
(b) there was a shortage of supply of the main brand which resulted in one depot running out of stock on Thursday and the other working on an allocation basis for the whole week;
(c) the shops were full of the main competitior's products because they had launched a pilot trial the previous week which had been a flop.

He would then wish to test out some ideas like:

(a) How long will the stocks of the competitor's trial last at current rates of consumption? Because product should be fed to the retailers to replace it at the right rate.
(b) Is the shortage of our product likely to interfere with the plan to feed product into the market at the required rate?
(c) Is weather likely to change and how?

In addition he would be likely to continue to ask questions of the kind that force an examination of strategic intentions, such as:

(d) Should I concern myself with this part of the country where the scene has been somewhat spoilt by the failed promotion of this competitor?

or

(e) Should I make sure that the competitor having failed in his trial will continue to find the market inhospitable?

After all, Tsun Tsu, the greatest of the known Chinese writers on war in ancient times, felt that one of the most effective means of overcoming an opponent was to attack his strategy.

Hard data cannot be the key factor in answering such questions as these. The fortitude of the competitor as well as the extent of his resources is likely to weigh heavily in the balance. The hunch about a competitor's vulnerability may come from an intuitive assessment of performance in the field. Hard data in support of such assessments is not easily obtained because it is likely to cut across the normal lines of analyses of sales and product performance.

Attempts to overcome these weaknesses in management information systems have tended to lead to rather complex and expensive procedures. Matrices of information may be prepared but they are expensive to maintain and sometimes complicated to interrogate. Time-series, or periodic data, is normally maintained for a fairly short time and even then is likely to have its basis changed from time to time so that any really extensive surveys of past performance are either unreliable or so costly to prepare that they are to be avoided.

CONSTRAINTS IMPOSED BY CURRENT INFORMATION SYSTEMS

Whilst some information systems are becoming more responsive and flexible, there is no doubt that the majority suffer from the narrow manner in which they have been devised. They are likely to have the following shortcomings:

(1) their basic orientation is inward-looking, reporting only on the performance of elements of the enterprise being managed, i.e. about sales, costs and profits;
(2) they do not look outwards at the aspects of the environment which affect performance, i.e. economic forces such as money supply, consumer spending or competitive activity;
(3) they do not relate trends in performance to changes in the environment so that the enterprise, and many of those managing it, can learn from their experience, i.e. analyses of failures to meet targets are not always objective.
(4) they do not provide any adequate basis for a meeting of the minds of those who operate predominantly from one or other hemisphere of the brain, i.e. the procedures devised by business systems designers are likely to be mechanistic in their approaches to problem-solving in a real world where the leaders may be mainly intuitive people;
(5) many of those making important decisions about operations of the enterprise do not look at the reports provided by management information systems even when they have useful information to impart. They tend to regard them as being unable to answer the most important questions.

Therefore we must regard the current state of the art, of using computers to help managers make their operating decisions, as being far from ideal. The opportunities provided by the vastly increased capacity of the microprocessor and its successors for handling all kinds of data should therefore be considered as offering possibilities of radical change which could provide for better support for the decision-maker in future.

MANAGEMENT DECISION-MAKING IN THE FUTURE

Several key factors are going to affect the resources available to management to help them make decisions in the future:

(1) most kinds of computing and data-handling will become abundant and cheap enough for individuals to have access to far greater amounts of information and calculating power than has been considered possible in the past, even for large enterprises;
(2) distance will not be an important factor in the cost of transmitting data from one place to another;
(3) it will be possible to store very large volumes of information in a relatively small space.

One of the consequences of these changes in the technology of computing will make it almost impossible for the present centralised system orientation to continue. The bargain between the manager and the operators of the large-scale system, in which he is asked to sacrifice his individual and idiosyncratic requirements to obtain a ration of standardised information, will no longer be attractive. This is already happening with the advent of the programmable calculator which offers reasonable power and flexibility at a fraction of the former cost of computing. A manager can currently work out trends, model costs, establish inventory re-order levels and order quantities without the need to employ anything but a pocket calculator. In future such small devices are likely to have facilities which make them even more powerful and hence the manager even more independent.

This independence of the manager will change the basic orientation of system design. To obtain his support designers will be obliged to create systems with a much more flexible response to users' needs. Those working mostly in areas where judgement is important will be able to insist that systems are designed to help confirm hunches and flashes of intuition rather than to churn out reports in a standard form which are useless for decision-making. Such reports as are required by auditors and those who are checking and maintaining the system will, of course, be made available. Otherwise the management reports produced are likely to be on request and to the specification of the individual manager concerned.

It follows that in future the manager will be less constrained to think like a primitive machine because his information system will no longer encourage him to do so. This will undoubtedly lead to more reliable and successful decision-making with greater emphasis on intuition and creative thinking.

A second result of the current thrust of computing technology will be to enable great concentrations of distilled information to be made on demand. This has already been done in many military command headquarters for a number of years; but using the previous technology, costs have been enormous. With the current expectations of technological advance in this field, it seems likely that many medium-sized enterprises will be able to provide their middle and top management with a comparable range of current information. This will cover all relevant information about the environment, its trends and its relationship with the current performance of the element of the enterprise under consideration.

The third result of the advancing technology will be to enable those using it to achieve a higher rate of performance improvement than has hitherto been possible. Response to changes in the environment will be faster and so too will the results of operational decisions be forthcoming and visible. Simulations of situations will be easily reproduced for training purposes and managers of enterprises will be able to engage in simulations of real-life business situations to a far greater degree than has hitherto been thought possible. One can visualise a planning room, rather similar to, though not as pretentious as, some of the military command posts seen in recent films. In such a planning room the computers could be switched to simulation and several years of operating decisions could be tested in hours. Alternative strategies could be tested against different competitive and environmental scenarios. When such activities could be closely associated with current operating data, their role could be seen as having both training and operating value contributing greatly to the rate of learning and improvement mentioned earlier.

When such facilities are seen not as the end-product of huge investments in hardware and software, but as the consequence of an assembly of microprocessor units of relatively modest cost, the implications are manifest.

THE CHALLENGE OF CHANGE

Undoubtedly there are adaptations to be made. But the manager will be in a stronger position in relation to the technology; thus in some cases being able to insist on better designs of systems. It is up to him to ensure that the initiative almost lost to mechanistic ways of thought in the first half of this century, is recovered fully by people who can think creatively, intuitively and objectively about decisions.

So there is a need to consider now how best to begin the process of transformation which will lead to our being able to take advantage of the new resources which are even now progressively being made available.

THE WAY AHEAD

The concepts we have been considering in this chapter will lead to a new kind of enterprise being evolved. The main task of these enterprises will be the assembly of intelligence or information combined with the control of the means of providing users with goods or services. They will be information intensive, providing themselves, and others in association, with the means of controlling and using expensive resources economically.

The need for such enterprises is already evident and in some cases the forerunners of these new types of information-based organisation are already operating even if they are in rather an unsophisticated state of development. The self-regulating aspects of such enterprises will be evident. For example, the ubiquitous radio taxi service is one in which the basic unit of operation is a car and driver. The radio station is the binding and controlling force despatching cars to customers and perhaps carrying out other functions also. There is a clear obligation on the part of the unit operators to the radio station operator and this makes the system stable but responsive under normal conditions.

The telephone-selling operations growing out of the increasing commercial use of the telephone network are another example of the information network of customer contacts which will probably be the nucleus of many an information-intensive enterprise.

The recent apparent diversification of the major clearing banks into the credit card business has had the effect of extending the influence of the banks through a fairly primitive system of cash substitution which has been progressively eliminating the use of cash in most consumer transactions in the USA for the past 20 years.

The growth of tourism has been dependent upon an information-intensive type of enterprise right from its beginnings. These bring into play resources of aircraft, buses and accommodation in widely dispersed locations. Those introducing computers have been able, quite early on, to progressively offer more and more individually tailored packages to suit the particular needs of consumers.

In every case mentioned so far the systems are potentially information-centred even with the hardware available in the past 5 years. All are developing progressively, like the tour operators, to offer more and more sophisticated services with the aid of better telecommunications and now microprocessors which are becoming available.

In this new era, command of resources through their capitalisation seems to be less significant than the command of the essential information service and control through research, promotion and interpretation of market influences. The vertically integrated organisation will perhaps be persuaded to allow its operating units to dance to the piped information that interprets market strategies into plans which can be fulfilled cheaply and effectively. The cobbler and his last seem, once again, to be potentially profitably joined.

This style of operation will be volatile, dynamic, responsive and idiosyncratic. It will provide opportunities for the creative, intuitive manager on a scale hitherto unsurpassed. It will also be the creator of many new small-scale enterprises with physical resources which the information- based enterprises will coordinate, with a new facility which will depend almost entirely on the cheap reliable services of microelectronics and microprocessors which are about to become abundantly available.

11 Responding to the Impact of Microelectronics

L. MARCH

Microelectronics has become a major concern of the management of most businesses because it has impact across the range of business activities, on the product cost and performance; on the manufacturing processes and on the information control and processing.

With such a wide range of impact across very diverse business sectors it is especially difficult to generalise on the nature and extent of the impact of microelectronics on an existing business. However, industrial experience to date has shown that it can drastically change the business environment in some very basic ways:

(1) more and different competitors;
(2) new product economics;
(3) shorter development and product life-cycles;
(4) new skills requirement.

Changes in the competitive scene arise from several possible sources – semiconductor companies integrating forward – for example, the move of these companies into calculators, watches, DP equipment, industrial controls, leisure products; traditional companies integrating backwards – for example, the domestic applicance manufacturers who are designing their own controllers and dealing directly with the semiconductor companies.

Similarly, there are examples of electronics companies and small private enterprises moving into markets with electronic-based products to challenge the traditional markets of mechanical and electromechanical manufacturers.

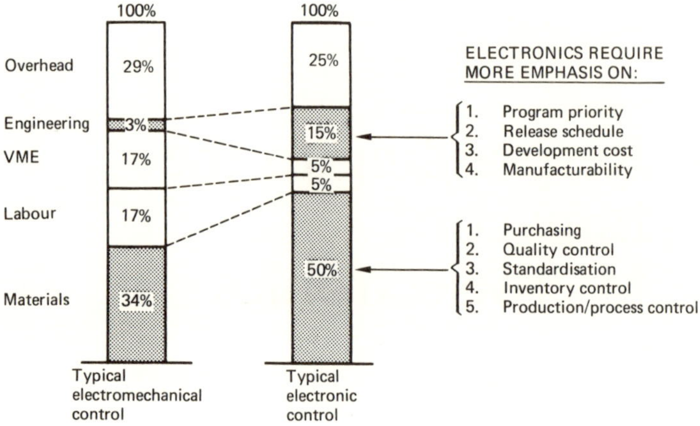

FIGURE 11.1 Comparison of typical product cost structures

The second basic change created by microelectronics is on the product economics. Figure 11.1 compares the cost structure of an electromechanical product and its electronic equivalent. The increase in engineering content and material costs are of special significance in the implied necessary emphasis and effort needed for the associated activities. Also of importance is the reduction in direct labour content, which can influence, along with other considerations discussed in previous chapters, the traditional processes of work-floor management and overhead allocations.

The third point – a reduction in product development cycles – can be one of the greatest problems as it can necessitate a more dynamic and reactive style of management. Figure 11.2, again based on electromechanical products, shows the considerably faster design cycle possible with electronic-based products – major advantage in a business that needs to react to market requirements and competitive activity.

Finally, microelectronics necessitate new skills throughout the organisation, not only in the design and engineering functions, where new tasks have to be undertaken – particularly software design as will be discussed later – but also in procurement now dealing with components exhibiting decreasing price trends; in sales and service; in manufacturing, where new processes will be needed; and in testing and quality assurance which may need to become more sophisticated and utilise automatic testing equipment.

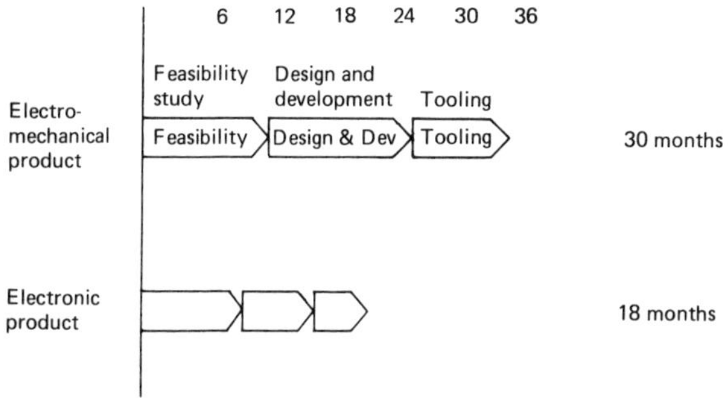

FIGURE 11.2 Typical development cycles for industrial controls

The sum effect of the changes that can result from moving into electronics can be very drastic. Table 11.1 lists some of the changes that have occurred in companies that have moved into electronics, and demonstrates not only that the impact is spread throughout the business, but also that the impact can be very great in areas remote from the design department and can, in fact, alter the entire character of a company. This is emphasised by Table 11.2 which compares the profile of a company before and after the transition to electronic products. The new-style company will have to be managed differently to allow it to survive in the more crowded competitive scene; to control its R&D, which represents an increased investment, and to ensure its ability to respond to the requirement for much faster product development cycles.

TABLE 11.1 *Possible impact of moving to microelectronic technology*

Area of impact	*Examples*
Cost structure	Labour content dropped from 35 % to 8 %
Manufacturing plant	Test equipment rose from 5 % to 30 % of plant
Skill requirement	Development staff – 65 % needed retraining
Key manufacturing areas	Procurement and material management identified as key areas
Service strategy	Throw-away strategy reduced service requirements by 25 % over 2 years
Product life cycles	From 8–10 years down to 3 years
Vulnerability	From 3 competitors to over 30

Table 11.2 *Impact on Company Profile of Moving to Microelectronics*

	Before transition (1968)	*After (1978)*
Primary technology	Spring and pendulum	Piezoelectric crystal and microprocessors
Value added by electronics	10 %	50 %
Product development time	2–3 years	6–12 months
Product life	Over 10 years	3–5 years
Operations	1 main plant	Decentralised
R&D investment	—	More than doubled
Market share	30 %	25 %
Competitors	5 major	Approximately 60
Profits	—	Increased

Microelectronics is deserving of special management attention, not only because of the potentially serious impact of the technology on business, but also because the process of selecting the most appropriate application areas, and managing their implementation, can present considerable challenges.

Successful implementation, that is effective development of those projects which provide greatest benefit to the business, can be achieved by adopting a three-part approach.

(1) identifying threats and opportunities that arise and assessing their overall attractiveness;
(2) assessing the resource implications of developing the preferred projects and choosing between the options available for providing those resources;
(3) managing the development to ensure effective utilisation of resources.

SELECTING THE RIGHT PROJECT

The selection of projects for microelectronic-based development should be such as to meet at least one of the following three objectives:

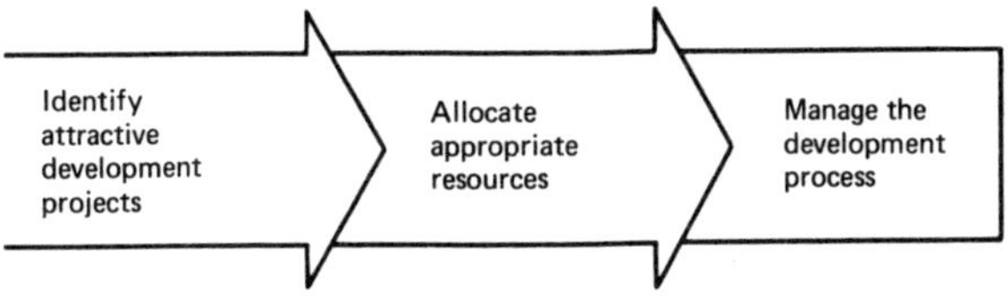

FIGURE 11.3 The three-stage approach to implementing microelectronic projects

(1) respond to market needs or opportunities;
(2) increase product attractiveness;
(3) improve company profitability.

Responding to market opportunities has been the objective behind the drive into electronic watches, calculators and games where a massive market was seen by the semiconductor manufacturers and where the traditional suppliers either failed to identify the threat or were unable to gain the technology in time to respond.

Figure 11.4 shows the situation of appliance control manufacturers in 1978, when they were faced with a forecast of rapid decline in the demand for their current product range and of all future growth forecast to be provided by electronic controls. For them the need to implement electronic controls was essential to retain market position.

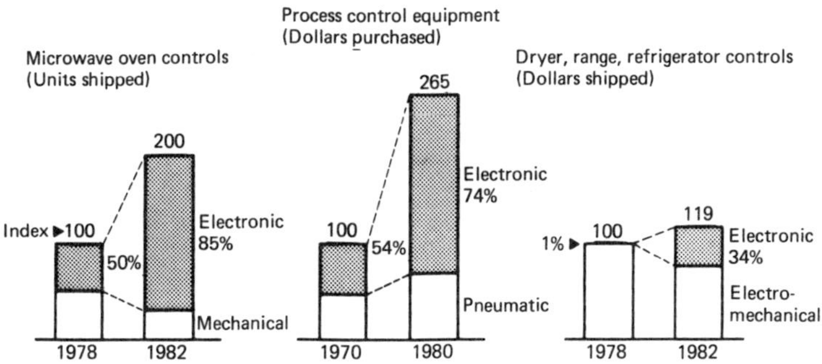

FIGURE 11.4 Impact of electronics on US industrial controls market

Increased product attractiveness can also be the objective in moving to electronics – possibly to counter market threats or alternatively to increase profit share and gain a competitive advantage.

It is very important, if this is the objective, to analyse the economic benefit of the user, and sedulously avoid the pitfalls of producing products that are necessarily complex and difficult to use, or which provide facilities that are not valued highly by the customer. A typical example of over-complex design for engineers can be found on some domestic appliances where the user is presented with a host of controls and displays that can by their complexity deter the user from gaining full benefit from that appliance. An analysis of five possible projects is shown in Figure 11.5, where they are compared for additional user benefit gained by incorporating processor technology in their design. Although all the projects may be practically feasible, only one provides clear user benefit.

Possible microelectronics application	Potential user benefit				
	Reduced cost	Improved reliability	Reduced size	Improved performance	New features
Project A					
B					
C					
D					
E					

FIGURE 11.5 Format for assessing user benefits

The third major consideration, that of improving company profitability, widens the scope of possible applications from those of product strategy considered above, to include internal processes both in manufacturing and administration.

Figure 11.6 illustrates in concept how this is achieved. Working from the RoI, the factors that contribute to it are broken down (usually to a greater detail than that illustrated) and applications are identified, based on the industry characteristics, that could contribute to an overall improvement in the return on investment ratio.

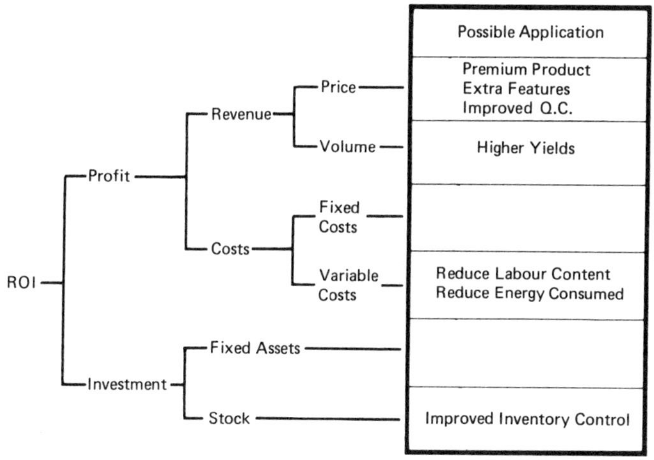

FIGURE 11.6 Analysis of impact of project on RoI

It is then necessary to review the identified opportunities and rank them in order of their leverage on the overall performance of the company. In Figure 11.7 the calculated leverage of four opportunities identified by an engineering company are compared. It would need considerable improvement in product component cost to make this opportunity more attractive than that of minimising the energy consumption.

The third stage of developing this analysis is to evaluate the possible improvements available from each project and factoring this by the

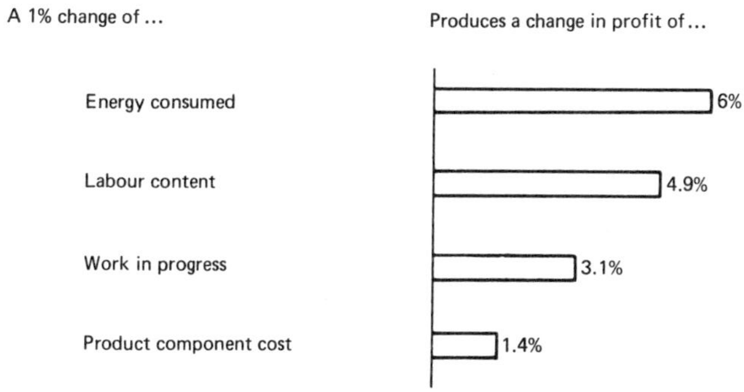

FIGURE 11.7 Analysis of leverage of possible projects

leverage to determine the overall impact on the profit and RoI.

This form of analysis has been applied in the food-manufacturing industry, a significant feature of which is the large leverage that arises from reduction in costs – often as high as 25:1. Inspection of the industry cost structure shows that materials count for 60 per cent of the costs and further analysis would give the breakdown of this 60 per cent to show a typical wastage of up to 5 per cent. Obviously by automating the mixing, filling and weighing processes the reduction of this wastage possibly by 1–2 per cent, could considerably boost profitability.

Analysis of the three possible objectives of initiating a microelectronics-based project – as outlined above – can identify the projects necessary to maintain market position or to improve company performance. This approach, which does not need a detailed knowledge of microelectronic technology, ensures that the scarce resources available for project development are utilised to maximum benefit.

ALLOCATING APPROPRIATE RESOURCES

Having chosen the projects which provide optimum benefit for the company the next stage for management is to identify the resources needed to implement the chosen projects and to decide how these resources are best acquired. Resources are needed to perform four tasks:

(1) assist in the selection of projects and to translate these projects into engineering specifications;
(2) design and test the necessary hardware (if there is any hardware content);
(3) design and test the necessary software;
(4) provide post-design support and maintenance.

It is very difficult to quantify the relative load of these tasks and hence to determine the size of necessary development resource. However some guidelines are possible based upon experience of designs across a range of industries:

(1) one hardware designer can usually support two or three software designers;
(2) one average software designer can produce between 3000 and 7500 lines of code per year but this depends on several factors:
    (a) experience, especially of the type of application,
    (b) clarity and stability of requirements;

(3) support and maintenance account for 20–40 per cent of the total development effort but can be much higher if many changes are needed or if the original development was inadequate.

These guideline figures are translated into graphic form in Figure 11.8, and show that software is the major resource load. This is an important characteristic of most, if not all, microelectronic developments. Indeed as the cost of processors has dropped software development has grown to be the dominent cost, as shown by Figure 11.9.

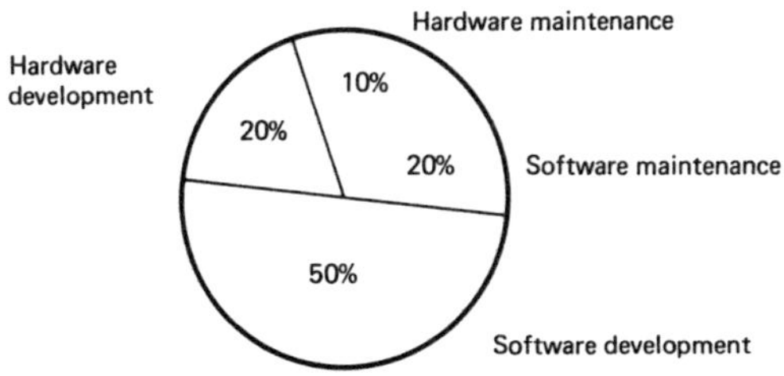

FIGURE 11.8 Allocating resources

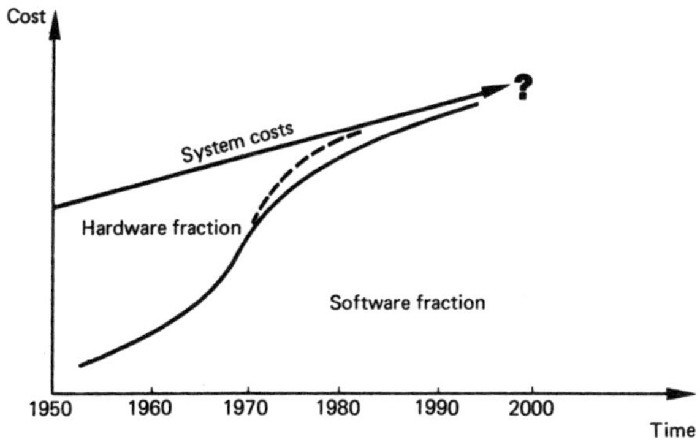

FIGURE 11.9 Growth of software as proportion of system costs

This is especially true in developments based on a bought-in microcomputer system or single-board computer, with which associated hardware development costs can be negligible.

The design equipment needed is, fortunately, much easier to define – for small- component-based projects it is a 'development system'. This is a microprocessor-based special-purpose device that provides facilities to both hardware and software designers to aid development, evaluation and testing of their project. These development systems are available both from the semiconductor companies to aid design based upon their own microprocessors, and also from electronic equipment manufacturers who usually provide facilities to support designs with a range of different manufacturers' microprocessors.

For larger-systems development, usually based on a bought-in processor, the availability of that processor is obviously desirable, but access to a larger computer with special software design aid programmes will improve productivity and considerably ease testing and error detection.

There are four options available to management for the provision of these resources needed for development of selected projects:

(1) re-train existing staff;
(2) recruit experienced staff;
(3) contract the design;
(4) acquire electronics company.

The selection between these options will be largely dictated by two factors:

(1) urgency of the projects;
(2) long-term importance of microelectronics to the business.

Retraining existing staff implies an external source of training which, however good, will still leave the necessity for development experience before reliable designs are produced and during this period the normal tasks of that staff will be neglected. This option is therefore preferred by companies adopting a 'toe in the water' approach who have not yet identified major threats or opportunities for their business.

Recruiting experienced staff implies a long-term commitment to micro-processor projects, but it is a slow and expensive process since experienced designers are in great demand and often require salaries above the normal scales for development staff. This approach is best for companies that have perceived a longer-term need for microprocessor technology but no urgency for immediate action.

The third option, that of subcontracting the design, does not remove the need for in-house expertise as this is frequently essential for selecting and controlling the contractor. However it does provide rapid access to a large and experienced resource such as may be needed to respond to a market threat. It also provides the ability to respond rapidly whilst internal capabilities are being developed.

The fourth option, that of acquiring an electronics company to provide the necessary resources and experience, is favoured by companies that have identified microelectronics as being of fundamental importance to their future and who are willing to make a significant investment to gain access to the technology. Unfortunately this approach is often less attractive than it appears:

(1) identification of candidates can prove a lengthy process;
(2) evaluation of their capability can require high level of expertise;
(3) competition from other companies also hoping to buy in technology can delay the process and push up the price.

These four options are summarised in Table 11.3 with an assessment of the *business situations to which they best apply.*

Managing the Development Process

The final stage, managing the development of the microelectronic project, deserves special attention – not because of the special difficulties and problems of this form of development but paradoxically because of the similarities to the development process in other engineering disciplines.

There is a mistaken belief that microprocessor development cannot be easily controlled and that normal development management processes do not apply. This has resulted in a tacit acceptance by some management of late delivery of projects, the performance of which does not meet expectations.

In fact the development of microprocessor-based projects – whether product-, process- or system-based – can be managed by a similar process to that of most other forms of development.

Many development managers have been frustrated by the delays that occur in completing the development of microprocessor projects seen in the early stages to be progressing well. Figure 11.10 shows such a project where progress appeared on schedule until it was 90 per cent complete

TABLE 11.3 *Summary of options for acquiring development resources*

Resource options	*Speed of response to threats and opportunities*	*Long-term commitment to microelectronics*	*Appropriate business situation*
Train existing staff	Low	Low	No immediate threats
Recruit experienced staff	Medium	Medium	Threat/opportunities identified but no rapid response needed
Contract design	High	Low	Rapid response to immediate need – or interim solution while internal resources are being acquired
Acquire electronics company	Medium-High	High	Forecast of major impact on industry by microelectronics

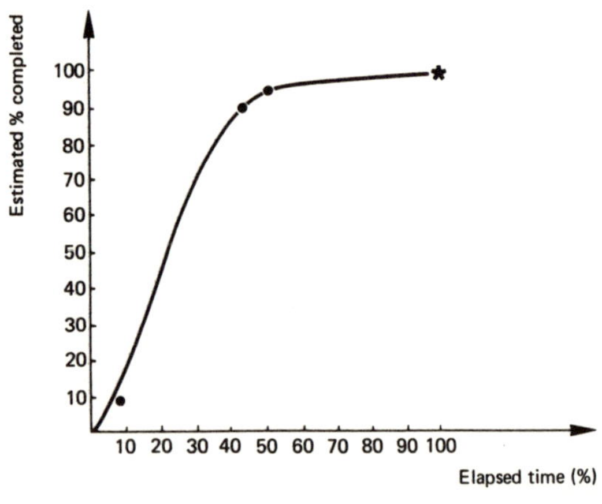

FIGURE 11.10 Analysis of system development status estimation

and then slowed down to virtually zero progress. Investigation of this type of problem shows that there are three contributing causes:

(1) unstable specifications supplied to the development department – necessitating frequent design changes;
(2) inadequate translation of input requirements into design specifications;
(3) premature moving of the development into implementation before system design is complete.

There is one very significant factor in the above three causes: that they all relate to the early stages of the development process.

The necessity to produce stable and accurate specifications and to give special attention to system design before implementation is recognised by all successful development teams. It is especially important for microprocessor-based developments for two reasons:

(1) It is the area where most problems occur:
    analysis of the problems executed on typical microprocessor development shows that 70 per cent of them arise during system design and only 30 per cent subsequently during the coding or programming phase.
(2) Early identification of these problems can dramatically reduce development costs:
    Figure 11.11 shows how the costs of correcting problems increases exponentially as the project moves from system design

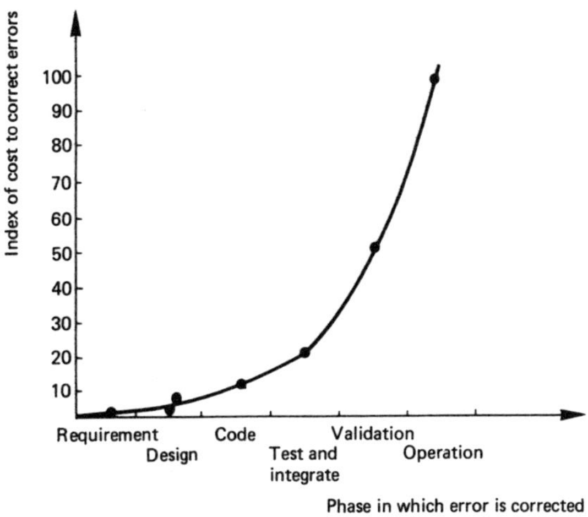

FIGURE 11.11 Cost of correcting software errors

> towards field application. Problems identified during system design can be corrected for 1 to 5 units of cost, but if left until field testing before identification the cost will exceed 100 units.

Clearly the avoidance of these front-end problems is vital for successful project development. Development management should therefore focus on three key tasks:

(1) to generate stable and accurate statements of requirements;
(2) to check that these are fully reflected by the engineering specifications;
(3) to ensure that the system design meets the requirements and specifications before major resource is committed to implementation.

Once the system design is complete the project progresses into implementation phase and subsequently into testing and validation. During this process the role of management is to ensure that adequate attention is paid to each phase and that it is performed adequately before moving into the next one.

The importance of system design, outlined above, necessitates a large proportion of the total project time being devoted to this activity – typically 35–40 per cent. This can prove very disturbing to senior management and users who see no progress in terms of output code or hardware design until way into the project. However the implementation phase is subsequently very rapid – typically only 16–20 per cent of the time – followed by a test and validation of 40–50 per cent.

Throughout the development process it is important to identify milestones against which the progress can be measured. There are fortunately easily defined and useful milestones – which considerably increase the visibility of development process. These are shown in Figure 11.12.

Microprocessor project development can be controlled focusing attention on the key areas: attention to system design, and the application of effectively applying a development process or processes so that management can assess status and progress against target.

As development experience with microprocessors has increased design techniques have evolved that simplify and improve the development task. Chief among these must be the structural approaches, including top-down design, structured programming and structural walk-throughs, which are described fully elsewhere, and which can

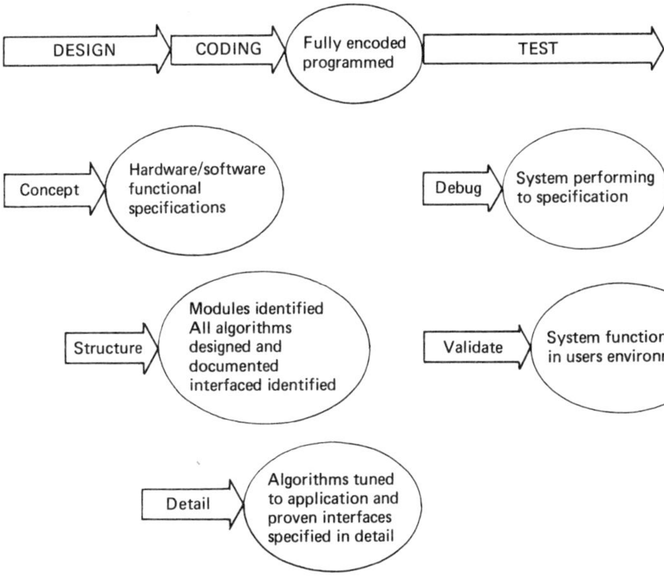

FIGURE 11.12 Milestones within project development process

remove a lot of uncertainty from the development process.

The impact of microprocessor technology is forecast to continue throughout the 1980s, affecting an increasing range of industries. This impact will be felt throughout those industries – by development, production, marketing and management.

The challenge to management is to anticipate the impact, to identify the opportunities and threats that it implies and to respond effectively by focusing resources into the key projects for their industry. The future success of their business could depend on how well they respond to the challenge.

Index